ENERGY EXPLAINED

ENERGY EXPLAINED

Volume 2:
Alternative Energy

VIKRAM JANARDHAN AND BOB FESMIRE

ROWMAN & LITTLEFIELD PUBLISHERS, INC.
Lanham • Boulder • New York • Toronto • Plymouth, UK

Published by Rowman & Littlefield Publishers, Inc.
A wholly owned subsidiary of The Rowman & Littlefield Publishing Group, Inc.
4501 Forbes Boulevard, Suite 200, Lanham, Maryland 20706
http://www.rowmanlittlefield.com

Estover Road, Plymouth PL6 7PY, United Kingdom

British Library Cataloguing in Publication Information Available

Library of Congress Cataloging-in-Publication Data

Janardhan, Vikram, 1971-
  Energy explained / Vikram Janardhan and Bob Fesmire.
      p. cm.
  Includes bibliographical references and index.
  ISBN 978-1-4422-0372-3 (cloth : alk. paper) —
  ISBN 978-1-4422-0374-7 (electronic)
  1. Power resources. 2. Petroleum—History. I. Fesmire, Bob, 1967- II. Title.
  TJ163.2.J34 2011
  333.79—dc22   2010008179

∞™ The paper used in this publication meets the minimum requirements of American National Standard for Information Sciences—Permanence of Paper for Printed Library Materials, ANSI/NISO Z39.48-1992.

Printed in the United States of America

To my wife Sandhya and our daughter Riya:
In writing this book, I have fulfilled a childhood dream of mine.
May you too follow your calling in all that you do.
—Vikram

To my wife, Gina, for making this book possible in so many ways,
and to James Mulkerin and Susan Morse (CHS 1980–1982)
for teaching me how to write.
—Bob

Contents
Volume 2: Alternative Energy

Part IV: Energy Efficiency   175

Part V: A New Energy Economy   209

Figures and Tables

Acronyms

AAA	American Automobile Association
AC	Alternating Current
ANWR	Arctic National Wildlife Refuge
AWEA	American Wind Energy Association
CAES	Compressed Air Energy Storage
CAFÉ	Corporate Average Fuel Economy
CARB	California Air Resources Board
CFL	Compact Fluorescent Lamp
CNG	Compressed Natural Gas
CSP	Concentrated Solar Power
CSR	Corporate Social Responsibility
DARPA	Defense Advanced Research Projects Agency
DC	Direct Current
DSIRE	Database of State Incentives for Renewables and Efficiency
DSM	Demand Side Management
DST	Daylight Saving Time
EEM	Energy Efficient Mortgage
EGS	Enhanced Geothermal Systems
EITI	Extractive Industries Transparency Initiative
EMF	Electromagnetic Field
EPA	Environmental Protection Agency
EPRI	Electric Power Research Institute
EV	Electric Vehicle
FACTS	Flexible AC Transmission System
FERC	Federal Energy Regulatory Commission

GNEP	Global Nuclear Energy Partnership
HVAC	Heating, Ventilating, and Air Conditioning
IEA	International Energy Agency
IPCC	Intergovernmental Panel on Climate Change
ISO	Independent System Operator
ITC	Investment Tax Credit
LEED	Leadership in Energy and Environmental Design
LNG	Liquefied Natural Gas
MPG	Miles Per Gallon
MTBE	Methyl Tertiary Butyl Ether
NERC	North American Electric Reliability Corporation
NHTSA	National Highway Traffic Safety Administration
NPT	Nuclear Non-Proliferation Treaty
NRC	Nuclear Regulatory Commission
NRDC	National Resources Defense Council
NREL	U.S. National Renewable Energy Laboratory
OPEC	Organization of Petroleum Exporting Countries
OTEC	Ocean Thermal Energy Conversion
PHEV	Plug-In Hybrid Electric Vehicle
PPA	Power Purchase Agreement
PTC	Production Tax Credit
PUC	Public Utility Commission
PV	Photovoltaic Cell
RES	Renewable Energy Standard
RPS	Renewable Portfolio Standard
SLOC	Sea Lines of Communication
SMES	Superconducting Magnetic Energy Storage
SPR	Strategic Petroleum Reserve
SVO	Simple Vegetable Oil
TOU	Time of Use
TRC	Tradable Renewable Certificate
TTB	U.S Alcohol and Tobacco Tax an Trade Bureau
UNCLOS	United Nations Convention on the Law of the Sea
UPS	Uninterruptible Power Supply
USCAP	U.S. Climate Action Partnership

RENEWABLE ENERGY

As a society, we have a fascination with weather. El Niño, Katrina, a killer heat wave—weather extremes capture headlines and the popular imagination. But aside from the spectacle of a natural disaster, there is an undeniable connection between humanity and the elements.

This relationship with nature—and in particular nature's energy sources—has existed throughout history. In many mythologies the sun symbolizes the chief god, who grants life to the other gods and to humans. People pray for rain or an early spring, and even today weather forecasts occupy a substantial part of the evening newscast, not to mention elevator small talk.

Old Mother Nature even dabbled in international affairs as far back as 1588 when the seemingly invincible Spanish Armada was defeated by a much smaller fleet of English ships thanks to a few well-placed storms that destroyed some of the Spanish vessels and scattered the rest.

More recently, human beings have learned how to capture nature's energy and put it to work. In essence, all energy sources are derived from nature and can even be considered to be "renewable." It's just that some renew themselves a bit faster than others. Wind is constantly being created whereas it takes quite a bit longer to turn dead plants into oil.

Volume I of this book examined nonrenewable energy sources, which for all practical considerations are limited in supply on earth. The chapters in this volume provide an overview of renewable sources of energy that are essentially inexhaustible.

What, then, is renewable energy? How is it different from energy derived from fossil fuels or nuclear reaction? The chapters in this part will provide a brief overview of the process of harnessing renewable energy and the impact of these sources on the environment.

Wind Energy   1

WIND POWER IS ACTUALLY a form of solar energy. About two percent of the sun's radiation that reaches Earth is converted to wind energy through the heating and cooling of the earth's surface.[1] When areas of the earth heat up, the warmer air rises and cooler air rushes in, creating wind.

The machines used to convert wind energy to electricity are called wind turbines or wind-turbine generators since they turn generators, not grinding wheels (i.e., windmills). Though often regarded as a different technology, the machinery inside a wind turbine is actually very much the same as that inside of a turbine fueled by natural gas or coal.[2]

A Brief History of Harnessing the Wind

Humans have used wind energy for thousands of years. It propelled boats along the Nile River as early as 5000 B.C., and by 200 B.C. was being used to pump water in China and grind grain in the Middle East. In America, wind provided electricity and pumped water on farms well into the twentieth century.[3]

Wind technology spread across the ancient world. European merchants and crusaders brought the technology back to Europe, and it continued to spread. The Dutch used it to drain lakes, and today are one of the world's leading wind technology exporters. Wind-driven pumps and mills came to the New World with European settlers in the nineteenth century.

Wind power experienced a decline with the industrial revolution as steam engines replaced wind in most applications. However, while industrialization obviated the need for small windmills, development of wind-driven generators continued.

Wind turbines as we know them today were first developed in Denmark near the end of the nineteenth century,[4] and by the 1940s an experimental 1.25 megawatt turbine was introduced in Vermont on a hill known as Grandpa's Knob.[5] This was a monster of a machine for the time (modern multi-megawatt turbines have only been in broad commercial use for around twenty years), and it supplied power to the local grid for several months.

More recent interest in wind energy has had its ups and downs, mostly attributable to the cost of traditional energy sources. The development of wind-turbine technology in the 1970s refined old ideas and introduced new ways of converting wind into useful power. The first wind farms, as we know them, sprouted up in the late 1970s and early 1980s. Today, wind turbine technology has matured to the point where these early installations look like scale models compared to the massive machines available now.

The Science and Technology of Wind Energy

The origin of wind is simple. The sun heats the earth unevenly, and where it's hot, air rises. Cooler air moves in to fill the void, and that movement is what we call *wind*.

Air doesn't weigh much, but it does have mass, and when that moving mass pushes against the blades of a wind turbine, it causes them to turn. The *hub*, where the blades come together, is connected to a shaft inside the wind turbine. Here, a series of gears known as a *gearbox* magnifies the speed of the rotation. The result is that the shaft exiting the gearbox turns at a much faster rate than the shaft coming in.

The high-speed shaft is connected to a generator that actually makes the electricity. The stronger the wind, the faster the turbine spins and the more electricity is produced. In the words of many a late-night TV commercial, it's just that easy!

From the generator, electricity moves through cables down the tower of the wind turbine and is collected by electrical equipment on the ground. From there, it passes through other cables and equipment until it reaches a transmission line and flows into the grid.

It is possible for there to be too much wind, in which case the blades and other components of a wind turbine can be damaged. If wind speeds get too high, the turbine has a brake that will keep the blades turning at a safe speed.

Wind turbines work best where the wind blows strongest and most consistently—on the tops of smooth hills, on open plains, near shorelines, and in mountain gaps. Offshore wind farms take advantage of the superior wind resources found at sea. To make electricity economically, wind speeds need to be in the 12 to 14 miles per hour range, again with the emphasis on consistency.[6]

Small wind turbines usually produce about 50 to 300 kilowatts of electricity. Larger turbines, often referred to as "utility-scale," range in size from 700 kW to 5 MW. However, as we'll get to in a minute, the variability of wind makes it a bit harder to equate these figures with the usual "number of homes served" measure.

Groups of wind turbines are often called *wind farms*, and they tend to be spread out over a large geographic area. Depending on the local terrain, one turbine needs about two acres of land, but the turbine itself takes up very little of that space, which can still be used for farming, grazing, recreation, and other uses.

Pinwheels and Propeller Heads: Types of Wind Turbines

OK, when you were a kid, did you ever blow into an unplugged house fan to see if you could get it to turn the other direction? Ever play with a pinwheel? If so, then you are already familiar with the most common type of wind turbine: the horizontal axis type.

Now another question: were you ever called a "propeller head" in school? Maybe you used the term to describe one of your brainier classmates? Well, whatever side of that particular fence you're on, you probably know where the term comes from. The beanie with the propeller on top—along with hummingbirds and helicopters—works on the same principle as the other main variety of wind turbines, the vertical axis type. These are also known as "eggbeater" style turbines since many of the designs resemble the business end of that classic kitchen appliance.

The vertical axis design is used mostly in small-scale applications, and is more likely to be seen on a rooftop than a mountaintop. Horizontal axis machines are more common, and this design is used exclusively in larger "utility-scale" turbines.

Vertical axis turbines are theoretically just as efficient as the horizontal type, but in practice certain physical stresses wreak havoc on what might look like a great design on paper. This is why vertical axis turbines tend

Figure 1.1 Two basic designs of wind turbines. (Doug Jones)

to be smaller. In addition, horizontal axis designs have a wider operating speed range and are self-starting.

The Challenge of Variability

The Rolling Stones summed up the fundamental difference between wind energy and other more traditional sources with their song, "You Can't Always Get What You Want." In this case, what you want is a nice, constant breeze.

Power grid operations are mainly focused on maintaining a constant balance between production and consumption, so the variable nature of wind power presents something of a challenge. However, there is already a good deal of variability in the grid because the amount of power being consumed is constantly changing. Studies and actual practice in places like Germany and Denmark[7, 8] have shown that a power system obtaining even as much as 20 percent of its power supply from wind farms can handle the inherent variability without any serious problems. In other words, wind can (and in some places, does) provide a major portion of a country's power supply.

Wind is also free, does not produce greenhouse gases or other pollutants, and is essentially infinite in supply. Given these advantages, it's not hard to see why wind power is being developed aggressively in many parts of the world.

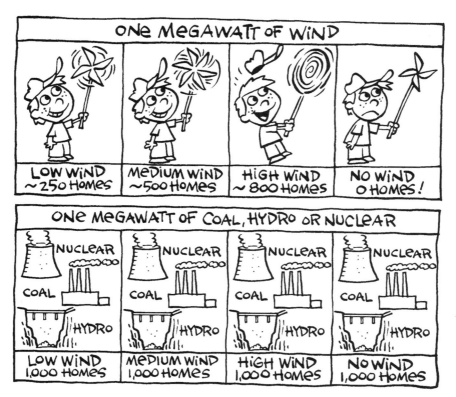

Figure 1.2 One megawatt of wind is not the same as
one megawatt of other types of generation. (Doug Jones)

Harnessing the Winds of Change:
The Economics of Wind Power

In the beginning, there was cost. And it was high. Retail costs for wind
power in the early 1980s were as high as 30 cents per kWh. As figure
1.3 shows, though, costs have come down dramatically, and during periods when natural gas prices are high wind is even cheaper than gas-fired
generation.

Wind is also inexpensive relative to other types of renewable energy
sources like solar energy, geothermal energy, or biomass, as shown in
figure 1.4. The up-front costs to build a wind farm are typically between
$1,000 and $2,000 per kilowatt of installed generation[9] (e.g., a 500 MW
wind farm would on the lower end cost $500 million to install). There are,
however, other costs associated with wind energy.

One of the primary hurdles that wind must overcome is the fact that
the best sites for wind farms are often far away from where the electricity

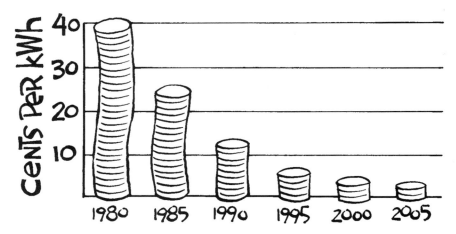

Figure 1.3 Cost of wind power in levelized cents/kWh. (Doug Jones)

would be used. Now this in itself is not a problem—we build transmission lines to carry electricity all the time. It is a problem when you consider that wind farms don't generate as much power in one place as say a coal plant. As we mentioned earlier, a wind farm often generates only a fraction of its rated capacity at any given time—but a transmission line has to be built large enough to carry the full rated capacity, so it doesn't overload during windy moments. That means the transmission line will be less than half full most hours of the year—not good economics for the owners of the transmission line! The people who put up the money for transmission lines generally like to recoup their investment as quickly as possible, and with wind, instant gratification typically isn't on the menu.

Ironically, there has been some discussion in recent years about spreading the cost of new transmission lines across both wind- and coal-generation facilities; for example, to exploit the coal resources of Wyoming's Powder River Basin and the superb wind resources of the Rocky Mountain ridges.

Let the Wind Blow: Market Forces Versus Government Intervention

Adam Smith, often regarded as the father of economics, refers to "the invisible hand" as the mechanism by which a benevolent God administers a universe in which human happiness is maximized. Down here on earth, however, there isn't a hand anywhere, divine or otherwise, that isn't

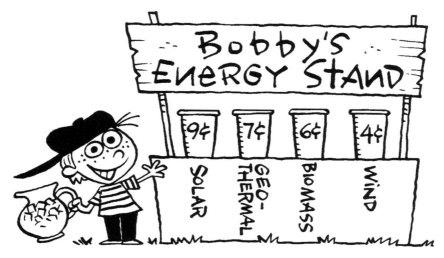

Figure 1.4 Wind costs compared with other renewable energy
technologies in cents/kWh. (AWEA/Doug Jones)

constrained by some form of government regulation. In short, the free
market is anything but free and that goes for the energy world too.

According to the United Nations Environment Programme, global
investments in wind, solar energy and other renewables totaled a whopping $223 billion in 2008, but this number would not be possible without
mandates and subsidies of various kinds.[10] Governments recognize the role
that wind energy can play in the energy mix. So, energy subsidies are the
not-so-invisible hands of government acting to shape the market in a particular way. In this case, the intervention is aimed at encouraging further
development of wind energy.

As we discussed in Volume 1, energy subsidies can be either direct or
indirect. Direct subsidies include payments from the government directly
to producers or consumers, or they can take the form of tax credits, which
work just like the tax credits individuals receive for, say, installing a more
efficient water heater or buying a home for the first time. The wind subsidy in particular is known as a production tax credit (PTC) because it is
only applied to energy produced by wind farms (and other renewables), as
opposed to an up-front subsidy for their construction.

Indirect subsidies cover pretty much everything except outright payments from the government. These include the provision of energy or energy services at preferential prices, funding for research and development,
or even providing insurance or loan guarantees.

How Do Wind Energy Subsidies Work in the United States?

The PTC was passed into law as part of the Energy Policy Act of 1992. This credit applies to the electricity produced during the first ten years of a wind plant's operation and is aimed at the development of large-scale wind energy projects.

Critics of wind energy often question the necessity or fairness of a tax credit that favors one particular technology. They see such mechanisms as tinkering with the free market. However, as noted earlier, the energy market has never been truly free.

Fossil fuel energy such as coal and oil have long benefited from government subsidies, usually of the indirect variety. For example, the federal government has paid out $35 billion over the past thirty years to cover the medical expenses of coal miners who suffered from black lung disease and has established legal protections to encourage the construction of nuclear power plants by limiting their liability in the event of an accident. Actions like these have a profound impact on the cost of doing business, and have been instrumental in maintaining energy prices.

Feed-In Tariff

At some point or the other, families quarrel about chores—whose turn it is to take the garbage out, mow the lawn, shop for groceries, and so on. This is more so the case in families that have children old enough to pitch in and help. Some parents foist the chores onto their kids, some parents nag incessantly hoping for compliance, while others coax, cajole, and even plead. And there is a group of them that use free market economics to get the job done—they pay their kids an allowance for doing their chores. They argue that in addition to getting the rooms cleaned, dishes loaded into the dishwasher, and meals prepared, it teaches their kids to manage money and learn to live within a budget. This does stir up debate among some as to why kids should be compensated for contributing to the general good of the family. And where should one draw the line with monetary incentives? Should kids only get paid for chores that contribute to the entire family (meal prep, lawn work) or should they get bribed for cleaning up after their own spills and messes?

Extend this analogy to include all of society, and you begin to have an idea about the topic of *feed-in tariffs*. A feed-in tariff is a government policy that guarantees producers of renewable energy a minimum price for selling power into the electricity grid. Put simply, the idea is to pay consumers and businesses for investing in renewable energy by setting the price the utility will pay them for the power they supply.

Feed-in tariffs are a form of an energy subsidy, but unlike tax credits that are a burden on all taxpayers, only electricity ratepayers (a.k.a. the utility's customers) pay for the feed-in tariff program. This is because the government forces the local electric utility to pay above-market rates to buy renewable energy, usually for a long period of time (15 to 25 years is typical). Undoubtedly, this raises the electricity bills for all customers serviced by that utility.

Germany and Spain are considered world leaders in implementing feed-in tariffs. Germany is not a particularly sunny place, but solar power has exploded there in large part thanks to feed-in tariffs. While other subsidies have enjoyed a long history in the U.S., feed-in tariffs have been largely ignored as a viable option here.

Critics, however, point to policy alternatives like renewable portfolio standards (RPS), and the fact that the increased electricity rates affect those with lower incomes the most severely. In addition, these programs have become so popular in some countries that their cost burden has forced the government to install a cap on payments. It's like a world where the hyperefficient kids do all of the household chores (and then some) giving their parents plenty of downtime to unwind and relax . . . but the allowances the kids collect begin to be a drag on the family's budget.

Wind Energy and the Environment

Perhaps one of the largest benefits of harnessing wind for electricity is the environmental impact. Wind turbines and wind farms do not emit anything harmful into the atmosphere. They do not produce hazardous liquid or solid waste. They do not deplete earth's natural resources such as coal, natural gas, or oil and, as a consequence, do not cause environmental damage due to extraction and transportation of these resources.

Wind energy has a lot of potential to offset the pollution caused by other kinds of power generation (especially coal) in the U.S. and worldwide. How extensive is that pollution?

Power plants in the U.S. emit about 70 percent of the sulfur dioxide, 33 percent of the nitrogen oxides (both contributors to acid rain), 34 percent of the carbon dioxide (a greenhouse gas), 28 percent of small particulate matter (a major cause of asthma and lung cancer) and 23 percent of the toxic heavy metals (which cause birth defects) that are released into our nation's environment each year. The displacement of fossil-fuel generation with wind power has a direct impact on all of these pollutants.

Naysayers to Wind Energy

Most all of us have witnessed one protest or another while channel surfing and some of you have undoubtedly been part of one. Protests against animal testing, against war, against people against war . . . you name it, there's probably someone somewhere protesting against it, and that includes wind power.

There are numerous benefits to utilizing wind energy—environmental and otherwise—but there are a few drawbacks. Wind critics point to bird and bat deaths, soil erosion at wind farms, and the need for backup generation. There is also a good deal of concern over wind farms' aesthetic impact on so-called "view sheds."

Upon further investigation, however, most of these criticisms can be dismissed. For example, while birds and bats do occasionally collide with wind turbines, the number of these animals killed each year by house cats far exceeds those killed by turbine blades. Soil erosion can be largely avoided if sound landscaping practices are observed.

With regard to backup power, it is true that a wind farm will require a certain amount of backup generation to even out the variations in its output. For a 100-megawatt wind farm, the backup required is about 2 megawatts.[11] However, that capacity already exists in the many conventional power plants currently generating electricity. It's also insignificant when you consider that people's use of electricity causes vastly greater variations in the balance of the grid than does the output of a few wind farms.

Aesthetic concerns are essentially subjective, and are, therefore, difficult to quantify. Noise from wind turbines has been greatly reduced as the technology has advanced and rules for minimum distances from homes have been implemented. A modern wind turbine placed 250 meters from your house would make about the same amount of noise as your refrigerator.

Visual impacts and reduced property values are what most residents object to when it comes to having a wind farm in their backyard, and for this there is no easy answer. Wind turbines located on land or in shallow coastal waters are likely to be visible to nearby residents (though distant offshore wind farms are now being built in Europe). Whether those residents see

Figure 1.5 Noise in decibels from wind turbines compared to other sources. (AWEA/Doug Jones)

**Fun Fact: Wind Energy Statistics
in the U.S. and Around the World**

- Worldwide wind generating capacity in 1999: 15,000 MW
- Worldwide wind generating capacity in 2008: 121,188 MW[12]
- U.S. wind generating capacity in 1996: 1,800 MW
- U.S. wind generating capacity in 2008: 25,170 MW
- Countries with largest share of power mix from wind: Denmark (23 percent), Spain (8 percent), Germany (6 percent)
- North Dakota has greater wind energy resources than all of Germany, but only 700 megawatts of wind generation is currently installed there[13] (as opposed to Germany, which in 2009 had 24,000 MW of installed wind generation).
- If the U.S. fully exploited its wind resources, the turbines could generate 11 trillion kilowatt-hours of electricity, or nearly three times the total amount produced from all energy sources nationally in 2005.

---

the turbines as a promise of a cleaner energy future or simply an eyesore will depend on the proverbial eye of the beholder.

Floating Wind Farms

As any sailor knows, winds at sea are often much stronger than those on land, and they get stronger the farther away from land you are. Recently, wind turbine manufacturers and power system planners have begun to explore the development of deep-water wind farms—large turbines located hundreds of miles from shore.

General Electric is currently developing turbines in the 5 to 7 megawatt range that could be situated on floating platforms. Researchers at MIT have done feasibility tests on these structures and have come up with workable designs that use gigantic cylinders for flotation and concrete for ballast. The platforms would then also be anchored to the seafloor.[14]

The potential for deep-water wind farms is enormous, and the European Union is already considering a "Super Grid" of massive wind turbines located in the North Sea. As high quality onshore wind resources are used up, and as an alternative to the siting battles associated with shallow-water installations, distant offshore wind farms could become a reality in the near future.

Airborne Wind Turbines

Picture a kite-like structure outfitted with wind turbines and flying at high altitude above the earth. Sounds like science fiction, but in fact is the subject of some serious inquiry. The idea would be to use a tether to transmit energy to the ground and keep the turbine-kite in place.

As you might expect, there are major impediments to the practical implementation of such designs. One sticky wicket is what to do if the wind dies down and the kite starts to fall back to earth. Another issue is keeping aircraft away from the tethers. Not surprisingly, no airborne wind turbines have been put into commercial operation as yet.

Solar Energy 2

*H*ERE COMES THE SUN might qualify as the most optimistic Beatles song of all time.[1] George Harrison wrote it in his friend Eric Clapton's garden, and it might have been a wishful exercise—in the months prior, Harrison had been arrested for marijuana possession, had his tonsils out, and quit the band at least temporarily. Whatever his motivation, the song could also serve today as a theme for the entire solar industry, which is enjoying a growth spurt like never before.

The sun has produced energy for billions of years—4.76 billion years, to be precise.[2] It burns at a temperature of nearly 29 million degrees Fahrenheit. Luckily for humans, earth is a little less than 93 million miles away from the sun—close enough to benefit from the energy it delivers without being cooked to a cinder. So, just how much energy does the sun deliver? Experts estimate that the energy from sunlight striking the earth for a period of 40 minutes is equivalent to our global energy consumption for an entire year.[3] Capturing that energy, of course, is another matter.

Solar energy arrives on earth in the form of solar radiation, which can then be converted into other forms of energy. We mostly use heat and electricity. Photovoltaic panels, or simply "PV," convert sunlight directly into electricity. These are what you'll find in rooftop residential systems, along with the cells that power calculators and remote call boxes on the highway. "Passive solar" is a term often used to refer to residential hot water systems or home heating that takes advantage of the sun but doesn't involve conversions to electricity.

Solar thermal plants—also known as *concentrated solar power* (CSP)— draw on both concepts. They concentrate the sun's rays to heat a fluid that is used to create steam that drives a generator. We'll get into more detail on both PV and solar thermal later in this chapter.

Fun Fact: Did You Know?

- The sun represents more than 99 percent of the mass of the entire solar system.[4]
- Biological rhythms that have periods of about 24 hours, also known as *circadian rhythms*, can be altered by long exposure to light. (Watch Al Pacino as detective Will Dormer in the movie *Insomnia* to get a sense for the chilling effects of altered circadian rhythms.)
- There is an entire field of biology that is devoted to the study of the impact of light on living organisms. It is called *photobiology*.
- *Photosynthesis* is the conversion of solar energy into biochemical energy in plants.
- Plants aren't particularly efficient—only about 6 percent of the energy they receive in sunlight is converted to energy they can use.
- Today's commercially available solar panels range in efficiency[5] from about 10 percent to 20 percent.

Harnessing the Sun's Energy: A Brief History

For centuries people have recognized the intimate relationship between the sun and life here on earth. In many religions, the sun is worshiped as a deity or the symbol of a deity. The sun is a Hindu deity called Surya. The sun god represents illumination and liberation. The Babylonians were sun worshipers, and the ancient Egyptians worshiped the sun god Ra. You might even say "sun worship" extends to the legions of oiled bodies tanning on the beaches of the French Riviera.

In fact, tapping into the sun's energy is not new. Humans have been doing it since the seventh century BC when magnifying glasses were used to make fire. The technology to utilize this energy, however, has improved dramatically and changed over the centuries. As early as 212 BC, our local star was reported to have played a role in settling international disputes when the Greek scientist Archimedes used the reflective properties of bronze shields to focus sunlight and set fire to Roman ships that were besieging Syracuse.

(The Greek navy re-created this experiment in 1973 and successfully set fire to a wooden boat at a distance of 50 meters. Students at MIT managed to set fire to a wooden mockup at 100 meters, but the TV show *Mythbusters* threw cold water on the idea by noting several reasons why, though technically feasible, it was highly unlikely ancient Greeks could have done it in practice.)

Chinese documents dating as far back as 20 AD, describe the use of mirrors to reflect sunlight to light torches for religious purposes. And for centuries, human civilization has designed dwellings with the specific intention of tapping into the sun's warmth.

The Evolution of Solar Technology: From Solar Cookers to PV Cells

The technology to capture solar energy made an evolutionary leap when Swiss scientist Horace de Saussure built the world's first solar collector in 1767. His device absorbed sunlight to cook food, and was later used by Sir John Herschel on an expedition to South Africa in the 1830s.

Another important event in the evolution of solar technology was the discovery of the photovoltaic effect. In 1839 a French scientist named Edmond Becquerel was experimenting with an electrolytic cell when he noticed that the generation of electricity increased when his setup was exposed to sunlight. Less than forty years later, William Grylls Adams and Richard Evans Day discovered solar cells made of selenium could convert sunlight directly into electricity. This discovery was hailed by Werner von Siemens as "scientifically of the most far-reaching importance," high praise from a guy who was regarded as being in the same league with Thomas Edison.

Discoveries like these soon led to other advances like the first solar water heater in 1891, the discovery of different photosensitive materials, and eventually the discovery that would change solar technology forever, the development of the silicon photovoltaic cell at Bell Labs in 1954. This was the first time that a solar device produced enough electricity to run an everyday appliance. Any dream of a bright solar future in the United States, where solar energy makes a measurable dent in our fossil fuel consumption, owes its origins back to this invention.

Additional Resources on the Web

History of Solar Cooking: A good description of Horace de Saussure's solar hot boxes of the 1700s can be found at http://solarcooking.org/saussure.htm.

History of Solar Technology: The U.S. Department of Energy has a very good website on the history of solar technology at http://www1.eere.energy.gov/solar/pdfs/solar_timeline.pdf.

History of Photovoltaic Cells and Solar Thermal Energy: http://www.californiasolarcenter.org/history_pv.html and http://www.californiasolarcenter.org/history_solarthermal.html.

What Is a Wavelength?

Light has many of the same characteristics as waves of water (which by the way carry energy too). Just as the crests of water waves can be close together or far apart, the crests of light waves can be close together or far apart. The wavelength is the distance it takes for a wave to repeat itself. For example, red light has a long wavelength but violet light has a much shorter wavelength.

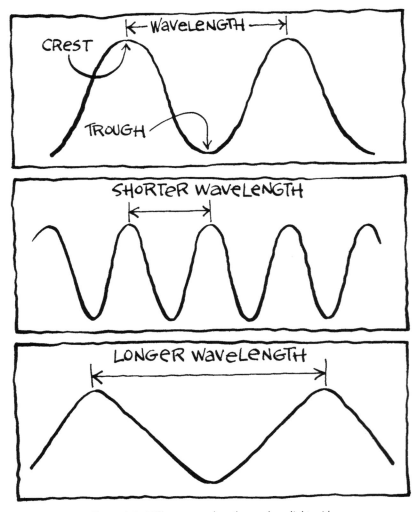

Figure 2.1 Different wavelengths produce light with different characteristics. (Doug Jones)

The Science and Technology of Solar Energy: Photovoltaics

A photovoltaic cell, commonly called a *solar cell*, is the technology used to convert solar energy directly into electrical power. Photovoltaic cells are typically made from silicon.

So how do they work?

Sunlight is actually made up of photons, which can be thought of as particles of solar energy. Photons are not all the same and they contain differing amounts of energy corresponding to the different wavelengths of the solar spectrum.

When photons strike a photovoltaic cell, one of three things happens. They might be reflected, pass right through the photovoltaic cell, or be absorbed into the cell. For the lucky (absorbed) photons that make it into the cell, what happens next is really kind of remarkable. The energy of the light coming into the cell is transferred to the semiconductor material (i.e., the silicon), which upsets the happy balance of electrons in the atoms that make up the material. Some of those electrons are shaken loose and they are then "directed" to an outgoing conductor by electrical fields.

The resulting flow of electrons is direct current (DC), just like what comes out of a battery. No fuel, no spinning machines. It doesn't even make any noise. The solar cell just sits there and generates electricity from light. Now really, admit it—that is pretty cool.

Individual photovoltaic cells only produce 1 or 2 watts, so many cells are electrically connected into a packaged weather-tight module that is in turn linked together with other modules to form an array.[6] PV arrays might be a few square feet to power, say a remote call box on the highway, or they might take up the entire roof of a big-box retail store.

Obviously, the output of a photovoltaic array is dependent upon how much sunlight hits the cells in the first place; so local conditions (e.g., clouds, fog, not to mention dust on the panels themselves) have a significant effect on performance. Given optimal conditions, though, most commercially available PV modules today come in at about 10 percent efficiency.

Yeah, that's not so great, but advanced designs have gotten as high as 40 percent efficiency in laboratory tests, and there are even some materials such as nanocrystal photovoltaics (discussed later in this chapter) that show promise of delivering up to 60 percent efficiency in the future.[7] The challenge, of course, is to make these higher-efficiency panels cheaper to produce.

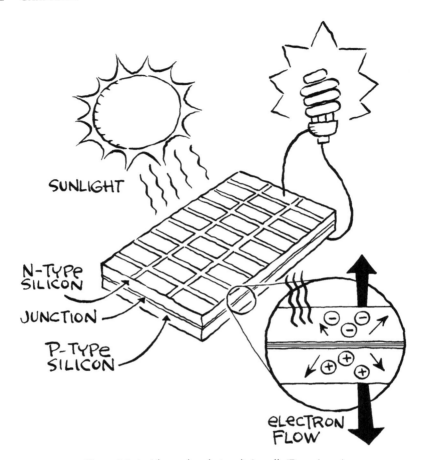

SUNLIGHT

N-TYPE
SILICON

JUNCTION

P-TYPE
SILICON

eLeCTRON
FLOW

Figure 2.2 Inside a solar photovoltaic cell. (Doug Jones)

One way solar efficiency has been improved is to use two or more layers of different materials with different abilities to absorb photons from the light spectrum. The material on the surface absorbs high-energy photons while allowing lower-energy photons to be absorbed by a second layer of another material. This technique of stacking many layers of cells on top of each other can produce much higher efficiencies. Such cells are called *multi-junction cells*.

Solar Efficiencies and Cost-per-Watt

Photovoltaic systems offer many advantages. PV systems are easy to install and don't require much in the way of maintenance. The environmental impact is minimal, requiring no water for system cooling and generating no by-products like ash or CO_2. So why, then, does solar energy only provide for barely 0.1 percent of our energy consumption needs in the United States?

There are two reasons for this: technology and cost.

Garden variety silicon is easy to find—just visit the beach and you'll be walking on it—but high-quality silicon required for the precision manufacturing of PV cells is in short supply these days. It already has a large market in the computer world (where do you think Silicon Valley got its name?), and now the PV industry is piling on.

The cost of a typical residential PV system runs into tens of thousands of dollars. The homeowner might make that back in the long run, depending on the local utility's rates and how long he or she stays in the house, but there's no getting around the fact that it's a lot of money to put up.

Research and development efforts are consequently being aimed at using cheaper materials. "Thin film" solar cells use printing technology to apply a layer of semiconductor material to a thin support panel, eliminating the need for silicon and drastically lowering the cost of production. Thin film cells are less efficient than traditional silicon cells, but dollar-for-dollar they are more cost-effective.

Boosting the efficiency of PV cells is the other main area of inquiry when it comes to making solar power more affordable. Converting a higher percentage of solar energy into electricity would mean that a solar panel of a given size would have higher power output, potentially reducing

---

Fun Fact: Did You Know?

- Each watt of photovoltaic power requires roughly 7 grams of silicon, which means that each 1,000-megawatt solar power plant needs 7,716 *tons* of processed silicon.[8]
- Photovoltaic cells were first used by NASA to provide power to satellites. The success of PV in outer space triggered the development of commercial applications for the technology.
- Photovoltaic cells, like batteries, generate direct current (DC), which is generally used for small loads such as electronic equipment but not for heavy loads like running a steel mill or a cement factory. For that, DC power is converted to AC using a device known as an *inverter*.
- A level of 40 percent efficiency is one of the highest achieved for any photovoltaic device in a laboratory environment.

the cost per watt. There have been some interesting advances on this front in recent years, but if you're waiting for a 40 percent efficient rooftop array to show up on the shelves at Home Depot, you'll probably have to wait a bit longer.

The Science and Technology of Solar Energy: Solar Thermal

OK, now we need to shift gears because this kind of solar power is completely different from photovoltaics. In fact, it's only "solar" in terms of what makes the heat—the rest of it is basically nineteenth-century technology.

In a solar thermal generator, the sun's energy is used to heat water directly or it is used to heat another fluid that in turn heats water through a heat exchanger. The resulting steam is then used to turn a generator, just like in a coal or nuclear plant. That's it.

Although seemingly less direct than PV, solar thermal is actually more efficient, converting roughly 30 percent of the sun's energy into electricity. However, one key advantage with solar thermal is that the heat can be stored to run the turbines well after the sun goes down. A large solar thermal plant in Spain does just that using molten salts as the storage medium, something we'll cover in a moment.

Solar thermal, also known as concentrated solar power (CSP), has been used in a fairly limited fashion so far, partly because it simply isn't practical for household use. There are a few utility-scale CSP plants in sunny places like Spain and the Mojave Desert in California.

Solar thermal comes in a couple of designs. In one, a patchwork of mirrors focuses the sun's rays onto a single central point, a collector often located in a tower. Water inside the collector is converted to steam and routed through a network of pipes to a steam turbine.

In another design, long trough-like mirrors aim the sunlight onto fluid flowing in a tube that runs parallel to the trough. The fluid, usually an oil, then flows into a heat exchanger where it turns water into steam to drive a turbine before being circulated back out to the mirrors again. This design is in use in the Spanish plant mentioned earlier.

Some experts have suggested that, in theory, a single gigantic solar thermal plant occupying an area of 10,000 square miles in the southwestern U.S. could provide enough power to serve the entire country (at least during the day). While this proposition is obviously beyond being

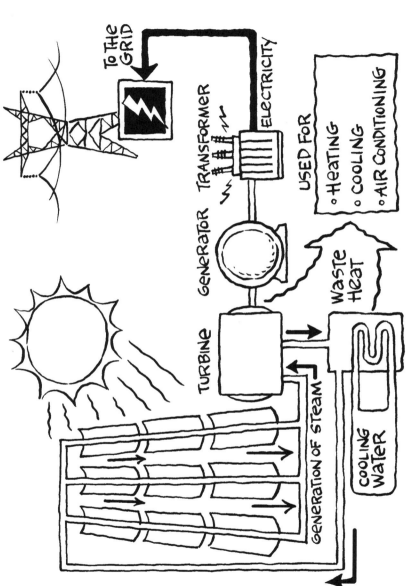

TO THE GRID

TRANSFORMER

ELECTRICITY

GENERATOR

USED FOR
- Heating
- Cooling
- Air Conditioning

WASTE HEAT

TURBINE

GENERATION OF STEAM

COOLING WATER

Figure 2.3 Solar thermal plants use the sun to run a steam turbine. (Doug Jones)

taken seriously, there is something to say for building several hundred solar thermal plants that collectively will take up just about one fourth of one percent of the land space in the United States to make a meaningful contribution to our goal of transitioning into an economy powered by sustainable energy sources. (The total land area of the U.S. is 3.79 million square miles, and 10,000 square miles of solar plants would make up 0.26 percent of the total area.)

Recently, there has also been some discussion about using solar thermal plants in the vast deserts of North Africa to generate power that would then be transported across the Mediterranean to Europe. There is even a consortium of companies pursuing the project, which has been dubbed DeserTec. (Full disclosure: Bob's employer is a member of the consortium.)

Energy Storage for Solar

Intermittency, variability—whatever you call it, solar suffers from the same affliction as wind power in that you can't rely on it to deliver a steady, predictable flow of electrons. Solar is a bit more predictable than wind, but the fact remains that to really make a dent in our energy portfolio solar needs some help.

We cover energy storage devices in the final chapter in this part, but there is one we thought we should mention here for a couple of reasons, and that is *thermal energy storage*. Turns out, heat is a great way to store energy, and solar thermal is a great way to produce it. So what's so great about heat? Well, for one thing you don't have to make the energy jump through a bunch of hoops by converting it from mechanical or solar energy into electrical and then storing it as an electric charge as you would with a turbine sending power to a battery. The energy captured in a solar thermal system just hangs around as heat until it is released to finish its trip to the turbine and generator.

How do you go about storing heat, you might ask? Doesn't everything just cool off? Well, yes, but some materials take a long time to do that and one of the slowest cooler-offers is salt, a handy substance we happen to have in abundance. Salt can be found in underground caverns (as we discovered in our discussions on oil), and of course there is a lot of it dissolved in ocean water.

When heated to a molten state, salt has the ability to store a great deal of heat long enough to be used after the sun goes down to make the steam that drives the turbine at a solar thermal plant. The technology is still relatively new and as you might guess it's still rather expensive, but the simplicity of capturing and storing heat energy avoids the losses associated with going from one form of energy to another and could make molten salt a key part of our energy future.

Chasing a Cheaper Kilowatt: The Economics of Solar Power

According to the Energy Information Administration, most parts of the U.S. have retail rates for electricity below 10 cents per kWh, but the cost can vary depending on the technology used to generate it. Coal and nuclear energy cost 3 to 5 cents per kilowatt hour. Natural gas can cost anywhere from 6 to 40 cents per kilowatt-hour depending on the whims of the natural gas market.

Solar thermal and PV cost around 15 to 17 cents a kilowatt-hour, so clearly for either one to become truly cost competitive, those numbers have to come down significantly.[9] As we noted earlier, much of PV's cost is tied up in all the silicon needed to make the panels. Still, even if alternatives can be developed, in most locations, solar power is still not ready for use on a mass basis if you're just looking at cost per kilowatt-hour.

That's not to say solar won't become cost-competitive with fossil fueled energy sources. Solar panels available when, say, Apple Inc. was moving out of Steve Jobs's garage (around 1980) turned out electricity at a cost of over $21 per watt. Today that figure is around $2.70 per watt, and it is dropping. Thin film solar panels are even cheaper, though as we've seen, they're also less efficient.

Solar technology is on the march, and like most every other technology, the further it marches, the cheaper it gets. We don't need to tell you how cheap computers have become relative to their capabilities, do we? Nah, we didn't think so.

But in addition to R&D there are other ways to bring down the overall cost of producing electricity from solar energy. Clustering solar power plants is one. Grouping multiple solar plants together, especially large ones, to share operational resources such as control rooms and spare parts could knock the prices down considerably. Another approach would be to simply build larger facilities to capture the sun's energy. One of the largest solar thermal plants in the U.S. is a 22-year-old installation in California's Mojave Desert.[10] It's capable of generating 354 megawatts of electricity. Local utilities are planning a 500 megawatt sized plant in the coming years. It's like buying toilet paper at Costco—the larger the plant, the cheaper the power.

Locating plants close to where the energy is used also brings down costs by eliminating the need to build long transmission lines. We probably won't be building multi-megawatt-sized solar thermal plants near cities anytime soon, but PV is another story. Solar panels can be placed just about anywhere, and they have the potential to virtually eliminate transmission costs (not to mention energy losses) altogether.

Comparing Costs of Electricity from Conventional and Renewable Sources

The cost of construction of a power plant is simply that—the cost of building materials, turbines, boilers, reactors, ancillary equipment, and everything else needed to create a power plant from scratch. In order to have an apples-to-apples comparison, the cost of building a power plant is divided by the amount of electricity it can produce. This gives us a dollar-per-kilowatt figure we can use to measure the cost of any form of generation. Coal plants, for example, cost about $1,300 per kW. Nuclear plants come in at more than $2,000. A solar thermal power plant can cost as much as $4,500 per kW.[11]

That's the cost of construction. The cost of *operating* a power plant is something different. It is a function of the fuel fed into the power plant, and other operating and maintenance costs incurred to produce a kilowatt-hour of electricity on a regular basis. The cost of operation also includes things like paying the personnel to run the power plant. This cost is expressed as dollars per kilowatt-hour.

Coal and nuclear energy cost 3 to 5 cents per kilowatt-hour. Natural gas can cost anywhere from 6 to 40 cents per kilowatt-hour depending on the price of natural gas at any given time (we've seen it range between $4.00 and $13.50 just during the writing of this book).[12] But what about renewable sources of energy? Aren't they essentially free? We don't need to pay to capture the wind or tap into the sun, so the cost of operating a renewable power plant should be 0 $/kWh, shouldn't it?

If you're talking about residential solar, the answer lies in the economics of displacement.

Let's assume you are considering installing solar panels on your roof. If you are paying your electric utility 5 cents per kWh, you don't care about how much it cost the utility to construct the power plant that generated your electricity. However, if you do install the panels, then in a sense you've become a utility yourself, and as such you do indeed care about the cost of construction (i.e., initial installation) of the solar system.

Installation costs are typically evaluated by computing a "payback period" that compares the cost of the solar-generated power with that from your utility. The difference between those figures allows you to arrive at a length of time required to recoup your initial investment.

That brings us to a classic business problem known as "make versus buy," and when it comes to solar the question comes down to this: do I make my own electricity with a solar array or buy it from my local utility? Probably the most often used yardstick in making that decision is the payback period, or simply the number of years it takes for the savings from avoided utility bills to pay for the cost of the solar system.

This sounds pretty straightforward, but beware—the assumptions you make about your energy needs, financing costs and especially utility rates can make a big difference. Then there is Uncle Sam, who currently is willing to cut you a break on your taxes to offset 30 percent of the cost of putting a solar system on your home. There's a good chance that deal could get sweeter, given the emphasis on moving away from coal and other fossil fuels toward renewable energy.

Our advice if you're thinking about solar for your home is to do your homework. Make sure you know what rates are available from your utility and if they are likely to change in the near future. Find out about state and local programs to encourage solar, too—a number of municipalities are starting to offer financing arrangements to bring the up-front costs of solar way down.

And be honest with yourself about how much juice your family is actually going to use. It is incorrect to assume that all homeowners looking to install a home solar system want to serve 100 percent of their total energy requirements with photovoltaics. It's fine to get a system that meets half or a third or a tenth of your energy requirements—and in fact, in many states (including California) it's much more economical (in either payback or return-on-investment terms) to just eliminate the "top tier" of electricity usage, since that's the highest cost per kWh.

In the end, though, most people who "go solar" do so for reasons that extend beyond simple economics. It's impossible to arrive at an ironclad payback period because many of the variables might change over time (hence the term "variable"). Get the facts, ask questions, and then decide what role the "intangibles" should play in your decision.

Solar Powering Your Home

Actor Brad Pitt and his Make It Right Foundation brought together leading architects and displaced residents of New Orleans's Lower 9th Ward following Hurricane Katrina. The foundation is funding green homes to replace those lost in the storm and the flooding that followed. A few of the houses were already built when Hurricane Gustav hit, testing the new buildings with 115 mph winds, but they all held up. Among the many green features, the homes sport solar PV systems on their roofs.

Beyond the feel-good publicity, though, solar power's appeal goes well beyond Hollywood celebrities. The advantage of solar power is that it is viable today both at residential and commercial scales. Now, we're not going to give you a step-by-step how-to guide here, but if you've thought about solar, we do have a few suggestions before you get started.

And just to be clear, when we say "solar thermal" in this section we're only talking about hot water systems, not power plants. Though, if you happen to live on a big plot in the desert and have a few million to spare in your bank account, by all means start shopping for trough mirrors and steam turbines.

Get an energy audit for your home.[13] Before you do anything else, it's important to get an energy audit done on your home so you can choose the right solar technology (PV, solar thermal, maybe both!). You can get an online audit done for free (see www.fypower.org/res/energyaudit/diy .html) or by calling your local electric utility company. The results of the audit will bear clues as to which solar technology is most appropriate to choose, not to mention other ways of cutting your energy expenses.

Identify the right solar technology for your home. We hope this book has provided you with a primer on the different solar technologies. A typical home is rated at about 3 kW of power. This translates to about 300 square feet of PV panels. For additional information about the different solar technologies, you can also visit the Energy Efficiency and Renewable Energy office of the U.S. Department of Energy at www.eere.energy.gov/solar.

Identify the right location for the solar panels on your rooftop. South-facing is best if you live in the northern hemisphere, but panels installed facing south, west, or east, angled between 5 and 30 degrees, can catch a healthy amount of rays nonetheless. Watch out for shade.

Get your permits and rebates in order. Solar systems usually require a building permit, or maybe a land-use permit if they're mounted on the ground. Your contractor should know what is required. Most states offer rebates for installing solar panels in your home. Good contractors handle this reimbursement process as well. To see what rebates governments and utilities are offering visit the North Carolina Solar Center and the Interstate Renewable Energy Council at http://www.dsireusa.org.

Get interconnection with the utility. As long as your utility company offers what's known as "net metering," you can plug your solar system into existing power lines, which take over your electricity supply at sundown and on rainy days when your rooftop cannot generate adequate electricity. Some states allow you to get paid for excess electricity that flows back into the grid. Typically your contractor will handle this as well.

Apply for your tax credits. Make sure you claim your credit on your federal income tax. Currently the benefit covers 30 percent of your out-of-pocket expenses (after rebates) for the solar system.

How Much Longer Are Solar Subsidies Necessary?

The U.S. and numerous other countries offer a variety of subsidies for residential solar power. Advocates of renewable energy say that this makes sense for a number of reasons like advancing the technology, creating a market to bring prices down and reducing the use of fossil fuels. In the U.S., renewable energy sources were subsidized with $4.8 billion of tax-payer money in 2007, more than any other type of energy source.[14]

What often gets overlooked, though, is that governments already subsidize the oil and gas industry to the tune of billions of dollars per year. It's just that the subsidies are indirect. They might take the form of a loan guarantee or favorable royalty payments to states with oil reserves, things that don't show up in analyses of direct subsidies.

Japan introduced incentives for solar energy in 1994 and has seen the average cost of solar energy drop by a whopping 72 percent due to the expanded market availability and increased efficiency of distribution. As a result, solar is now cost-competitive in Japan, and the need for incentives is shrinking. Germany is following a similar path and has ambitions of relying solely on renewable energy sometime in the next 15 to 20 years.

American incentive programs are increasing, too. In 2008, Congress passed historic legislation that extends the 30 percent federal investment tax credit for both residential and commercial solar installations for eight years. California's Million Solar Roofs initiative is one example of a grow-ing number of solar energy tax breaks being legislated at the state level. Municipalities are even getting into the act, creating financing mechanisms to allow people to lease solar systems and pay for them via their property taxes.

So are subsidies really necessary? Advocates insist that tax breaks and subsidies are necessary to help fledgling technologies gain a foothold in the industry. And if fossil-fuel based technologies get subsidies, the reasoning goes, then it is necessary for renewable sources to get them as well in order to ensure a level playing field.

On the other hand, renewable energy technologies do not *have* to receive government subsidies merely because the oil and gas industry has. It's possible that solar could become cost-competitive simply via "organic" growth. Prices of photovoltaics are going down every year, and the cost of standard electricity is going up. If these trends continue, there will be a meeting point where solar power has the ability to compete against fossil fuels without subsidies.

Additional Resources on the Web

You can read about state and local incentives on the Database of State Incentives for Renewables and Efficiency (http://www.dsireusa.org/). As mentioned on their website, DSIRE is a comprehensive source of information on state, local, utility, and federal incentives that promote renewable energy and energy efficiency.

The Energy Information Administration (http://tonto.eia.doe.gov/energy_in_brief/energy_subsidies.cfm) posts very useful information on various forms of government subsidies for energy. The report has details on the amount of subsidies per unit of production as well.

The Future's So Bright, I Gotta Wear Shades: The Future of Solar Technology

There are many reasons why one can look forward to a brighter solar future. We'll showcase a select few up-and-coming technologies that will give you a sneak peak at what's next in sun power. Of all the technologies for generating electricity, solar energy does share some interesting characteristics with the semiconductor industry that provide the potential for rapid and sustained increased output at decreased costs. If these technologies make their way from the research lab to mass production, you might soon be wearing solar cell coated baseball caps to the ball game and charge your iPod and cell phone on the fly. So let's take a closer look at the next generation of solar solutions.

Next Generation PV Cells: From Plastics to Nanotechnology to Flexible Solar Sheets

Crystalline-silicon solar cells are about 20 to 25 percent efficient at best; the ones deployed today in most rooftop systems are typically closer to 10 to 12 percent efficient. Beyond silicon, researchers are looking at various other materials to create solar cells.

Organic solar cells are based on flexible plastic and can be manufactured cheaply. They can even be painted onto a surface. They're not very efficient (only 3 percent), plus the organic material tends to break down in the sun, which is kind of a deal breaker.[15]

Other research is ongoing to look into a number of new technologies, such as lenses, chemical dyes, multilayer cells and so-called "quantum dots" that trap more of the incoming sunlight.

One such technology is called dye-sensitized solar cells or *Grätzel cells* (named after the inventor Michael Grätzel). Like thin film technology, these can be used in flexible sheets or coatings. Imagine a day when your solar cell–coated baseball cap powers the iPod and cell phone you carry, or the skin of a building powers the ventilation system. The main drawback is, once again, efficiency. Grätzel cells don't extract much energy from sunlight and their performance drops after a relatively short time.

Another promising PV cell technology already in use is thin film solar panels. The U.S. National Renewable Energy Laboratory (NREL) is working on new thin film solar technology that is close to the silicon-based variety in terms of performance.[16] These thin film solar cells may one day be draped around everyday items like caps, blankets, roofs of cars, and so on. In large-scale production, thin film cells can be significantly cheaper per watt to produce than current crystalline-silicon cells.

Multicrystalline silicon-based PV cells have reached 20.3 percent efficiency in the laboratory, but one commercially used thin-film technology known as CIGS has nearly matched that number at 19.9 percent.

Keep in mind, too, that even a fraction of a percent makes a big difference over the decades of a panel's useful life. Currently, silicon PV cells account for around 95 percent of the total solar market, but thin film is projected to take 20 percent (or more) of the total in just a few years.

NREL researchers have also shown that nanotechnology may greatly increase the amount of electricity produced by solar cells. An NREL team recently found that tiny "nanocrystals," also known as *quantum dots* (clearly, the better name for marketing purposes), produce as many as three electrons from one high-energy photon of sunlight. With current technology, sunlight is converted to at most one electron per photon. The rest is lost as heat. Quantum dots, in theory, could achieve efficiency levels of more than 65 percent, putting PV on par with even the most efficient conventional generation methods.

Solar Technologies for the American GI

U.S. soldiers' packs weigh around 100 pounds but a good 20 pounds of that is used up by batteries. To lighten this burden, the Defense Advanced Research Projects Agency or DARPA has set a goal for researchers and industry to invent a solar cell technology that delivers 50 percent efficiency in energy conversion. The reduced weight is an obvious improvement, but when coupled with solar rechargers, the difference in combat could be huge as equipment like night vision goggles, GPS, and radios could be used for longer periods of time.

In the lab, 40 percent efficiency in converting sunlight to electricity is considered the number to beat. DARPA's objective, then, is quite a tall order when you consider that improvements of even one percent represent a major advance.

Fun Fact: Recycling Silicon for Fun and Profit

Researchers at IBM recently came up with a process to harvest scrap silicon from the computer industry and turn it into solar panels. The process won the Most Valuable Pollution Prevention Award in 2007 from the National Pollution Prevention Roundtable.

Conclusion

Without a doubt, there are significant hurdles that solar energy must overcome before it can be expected to compete at scale with traditional forms of energy. Boosting efficiency, bringing down cost, and overcoming variability are the big three. But there are exciting prospects on all of these fronts.

Today, solar power is no longer a laboratory curiosity or a vanity project for hippie engineers. It is for real, and along with wind it represents the most likely first step toward a system of energy production that can survive in a carbon-constrained world.

Tidal Energy

3

THE SIGNS OF THE ZODIAC are broken down into a few basic elements: fire, earth, air, and water. Ancient cultures used a similar organization of elements to explain natural phenomena. The Greeks added "ether" to the mix, but the idea is the same: that everything in the world is made up of some combination of these primary elements. This concept persisted until the Enlightenment when the birth of modern science paved the way to a clearer understanding of nature.

In previous chapters, we discussed tapping into energy from "fire" (solar energy) and from "air" (wind energy). In this chapter, we will discuss energy from water, or ocean tides, to be exact.

Energy from the Ocean

We're all familiar with tides, but what causes them? If you said the moon, you're only partially right. Generally speaking, gravity causes tides but the gravitational forces come not just from the moon but the earth and the sun as well.

Let's start with the moon. As its relative position to a given beach changes over the course of the day, the moon literally pulls the ocean water toward it. So, when the moon is directly overhead, its gravitational force is strongest and the result is a high tide. Sounds simple, but what's interesting is that the same thing is happening on the other side of the planet, except that it's the earth's own gravitational force that is pulling the water. This is why in most places there are two high and two low tides each day, one primarily driven by the earth and the other by the moon.

The sun is the wildcard. Even with its massive gravitational pull, the fact that it is 92 million miles away means that it exerts just 40 percent as much pull as the moon, which is right in our celestial backyard at 233,000 miles away. Depending on its relative position in the sky, the sun can either amplify or mitigate tides. If, for example, the sun lines up behind the moon, you get the force of both bodies working together and the result is an especially high tide.

The same is true if the sun is directly on the other side of the planet from the moon, in which case it augments the pull of the earth. If, on the other hand, the sun is at a right angle to the moon in the sky, it results in a "neap tide," which is milder since the sun is partially canceling out the action of the moon and the earth.

So, how much of an effect do tides have? Well, like most everything, it depends on where you are. In the middle of the ocean, tides can change the height of the ocean by up to 40 feet (not that you'd notice). At the shoreline the change can be even greater in areas like bays where the flow of water is restricted.

"Ocean's Five"

OK, what if George Clooney and Brad Pitt only had three guys to help them take down a trio of Las Vegas casinos? Well, they probably wouldn't have been able to pull it off, and the film *Ocean's Eleven* wouldn't have been nearly as much fun. But we humans have only come up with five technologies so far for tapping into the energy of the sea, so the title of this section is by necessity well short of a direct movie tie-in.

From our earlier discussion of hydropower, recall just how forceful moving water is. The same physics are at work in tidal power, just in less obvious ways and with a few extra challenges to overcome.

Tidal power can be classified into two main categories.[1] *Tidal stream systems* work a lot like wind turbines except that instead of air, the stuff moving the turbines is water. These systems have a relatively light environmental footprint and are relatively inexpensive compared to other ocean power technologies. *Tidal barrages* still use moving water, but instead of using a constant flow they take advantage of the difference between high and low tides.[2] They work more like dams, except they generate power regardless of which direction the water is flowing. As water flows toward shore during high tide, it passes through turbines, or is used to push air through a pipe to turn a turbine. The same thing happens when water flows out at low tide.

Constructing a "dam" of this sort across a tidal estuary is rather challenging, and consequently very expensive. Tidal barrages are also only effective when the water is flowing in or out (i.e., about 10 hours a day),

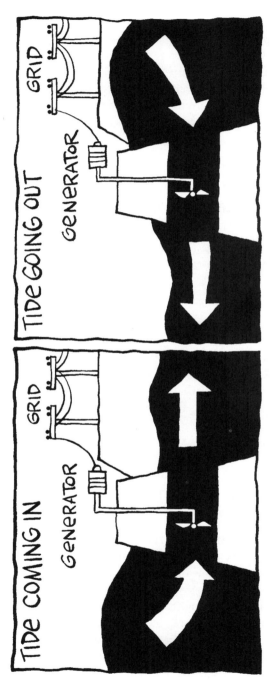

Figure 3.1 Tidal barrage. (Doug Jones)

so it's not very practical as a constant source of power. On the other hand, tides are highly predictable, so there's none of the variability that we associate with wind or solar.

Tidal barrages, however, come with some serious environmental drawbacks. They can alter water levels in the surrounding inland areas and they can act as giant washing machines, churning up the water to the detriment of sea life. There are only two large barrages operating today, one in La Rance, France, and the other in Nova Scotia, Canada. Sites for new barrages exist, but the U.S. is notably short on good ones.

Fun Fact: Did You Know?

Researchers in France estimate that if tidal power were implemented on a wide scale, the interference with natural water flows would slow the earth's rotation by 24 hours every 2,000 years. (Does that mean we wouldn't get old as fast?)

Tidal Fences

Tidal fences work essentially the same way as barrages, except that they completely block the channel. The environmental implications of such a system are, shall we say, less than ideal, so tidal fences are not likely to be built across rivers. However, the use of them in channels between land masses is being considered.

A tidal fence uses vertical axis turbines, much like the "propeller head" type of wind turbines we covered in chapter 1. In this case, the turbines are mounted inside a framework that spans a flow of water, as depicted in figure 3.2.

Tidal Turbines or Offshore Turbines

Tidal turbines are a new technology that can be used in many tidal areas, and you can think of them as underwater wind farms because the technology is essentially the same. None have been built on a commercial scale yet. Since water is roughly eight hundred times denser than air, it has the potential to generate much more electricity, but at the same time it puts far more stress on the turbine's machinery. That means tidal turbines will need to be engineered to a much high degree of strength than their lightweight counterparts, and yes, it means they will be more expensive.

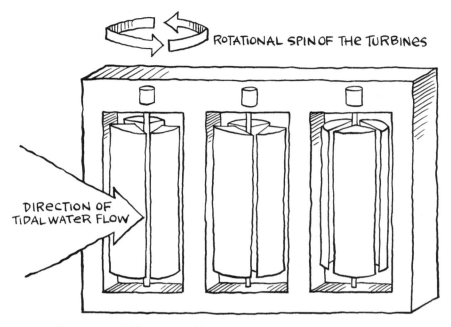

Figure 3.2 Tidal fences resemble vertical axis wind turbines. (Doug Jones)

Wave Energy

Ocean waves contain a tremendous amount of energy. Even a little two-foot roller can easily knock you over (we're speaking from experience here). Though often confused with power generated from tides and ocean currents, *wave power* is distinct and different.

Waves are generated by wind blowing over the surface of the ocean, so as long as we have wind and oceans, we'll have waves. It stands to reason, then, if this energy could be captured somehow, it would represent a reliable and inexhaustible source of power. In fact, the total energy of the world's waves has been estimated at between 2 and 3 million megawatts.

Wave power is still in a developmental stage, mainly because capturing wave energy is rather tricky. You can derive power from waves in one of several ways: directly, by focusing the water itself to drive a turbine; indirectly, by using waves to fill a reservoir and then releasing the water through the same kind of turbines used in hydropower dams; or by having floating devices trap the back-and-forth motion of the waves and convert it to other useful forms of energy.

Figure 3.3 The physics of one type of wave power device. (Doug Jones)

The first kind is known as the *oscillating water column* method and works by using a column of water as a piston to pump air and drive a turbine to generate power. This type of device can be anchored to the seabed or it can be built on shore. This technology is still in its infancy, but is being advanced in places like Japan where native energy sources are desperately needed to reduce the country's reliance on imported fuels.

Floating buoys are also used to capture wave energy and convert it to other forms of energy. When an object bobs up and down in the water, it moves around in the shape of an ellipse. The buoy can be designed to capture the energy produced by this movement.

The third kind of wave power machine is a *hinged contour device*, which follows the motion of the waves and creates power by harnessing the motion at the joints. It looks a lot like an accordion or multi-fold closet door.

Ocean Thermal Energy Conversion (OTEC)

OK, so far all of our oceanic power sources have involved capturing the energy of the water itself. But the sea offers another energy-making prospect by exploiting the difference in temperature between warm water at the surface and the much colder water deeper down.

Ocean thermal energy conversion or "OTEC" comes in two varieties: open systems and closed systems. We'll start with the second one since it's very similar to something we're all familiar with, the refrigerator. A closed OTEC system is basically a fridge working in reverse. A fluid with a very low boiling point, like ammonia, is pumped through a system of pipes at the ocean surface where it warms up, boils and the resulting gas is used to drive a turbine. The gas exiting the turbine flows down to colder depths where it condenses back into liquid form and the process repeats.

Obviously, this type of setup requires a certain amount of energy to run the pump, but the system as a whole produces a surplus of energy that can be sent back to shore via power cables. Understandably, there are practical considerations to be addressed (what if a whale bumps your generator?) but this setup would work in theory.

An open OTEC system takes advantage of ocean water temperatures, but without the use of a working fluid like the ammonia. The fluid in this case is the water itself, which is pumped into a vacuum chamber. The very low pressure inside the chamber allows the water to boil at a much lower temperature and the resulting steam is used to drive a turbine.

Now for the bonus—the steam coming out of the turbine condenses back into water, except now it is free of the salt and other materials it came

Figure 3.4 Wave energy in this design is captured where the floating panels meet. (Doug Jones)

in with. So, an open OTEC system produces not only power but also fresh water.

Hybrid systems that combine both open and closed system technologies are also possible, but so far there have only been a few OTEC prototypes built.

Fun Fact: Tidal and Wave Energy

- A tidal power study conducted in the 1980s estimated the world's potential ocean energy at over 330,000 MW.[3]
- Southeast Asia could be a hotbed of ocean power thanks to a large number of straits and narrow passages between islands and larger landmasses.[4]
- Tidal power technology isn't rocket science, but it's still expensive (where have we heard this before?) and the only large tidal plant working today is a 240-megawatt installation at the mouth of the La Rance river estuary in France. It's been there since 1966.

Conclusion

Ocean power is still being developed, but it holds a lot of promise. The various technologies we've covered are all renewable and sustainable and they produce no pollution (with the possible exception of a leak of fluid from an OTEC facility). There are even potential benefits (e.g., production of fresh water) that extend beyond energy. Research is continuing, and a new generation of ocean power devices is becoming increasingly cost effective.

Biomass Energy   4

MOST PEOPLE THINK OF MICHAEL JORDAN as not only the greatest basketball player of all time, but also a genuinely nice guy. During his professional career, off the court, he was soft spoken and cut an elegant figure. On the court, however, it was a different story. Jordan was reportedly a master of trash talk, subjecting his opponents to a steady stream of verbal abuse that contrasted sharply with his clean-cut image.

No one has yet figured out a way to convert that kind of trash into energy, but the more conventional kind is increasingly being used to do just that. And we Americans produce a lot of trash. The EPA estimates that the U.S. generates 200 million tons of waste every year,[1] most of which winds up in landfills, recycling programs notwithstanding.

Organic waste is biodegradable and has the ability to break down safely and essentially disappear into the environment as carbon, hydrogen, oxygen, and so on. Biodegradable wastes are 100 percent recyclable and comprise around 67 percent of the solid waste in a home (yeah, we were surprised at that number too, but it's for real).

One Man's Trash Is Another Man's Treasure: Introduction to Biomass Energy

The term "biomass" refers to organic material that essentially represents a reservoir of trapped solar energy. Plants, for example, absorb the sun's energy and that energy is passed on to animals and people when they eat plants. Turning biomass into energy is pretty straightforward: You burn it.

Yep, it's that simple. The material burns, producing heat that is used to make steam and . . . you know the rest by now, don't you? Steam, turbine, electricity? Yes, like so many other types of generation it's really all the same once you make enough heat to boil water. Biomass is considered

---

Fun Fact: Trash[2]

- America leads the world in waste production at 4.5 pounds of solid waste per person per day; the average American will produce over 112,000 pounds over the course of his or her life.
- The United States comprises 6 percent of the world's population and produces 40 percent of its waste.
- One third of all garbage thrown away by Americans is packaging.
- Between Thanksgiving and the New Year, Americans throw out up to 25 percent more trash, amounting to a massive five million tons of additional refuse. The vast majority, about four millions tons, is believed to be made up of wrapping paper and shopping bags.
- A disposable diaper takes 550 years to decompose, and an aluminum can takes 200 to 500 years while paper bags break down in about a month.
- 84 percent of all household trash can be recycled.
- Every ton of paper recycled saves 17 trees and avoids the use of 7,000 gallons of chemicals.

Burying 15,000 tons of trash in landfills creates one new job, while recycling the same amount of trash creates 9 new jobs.

---

renewable since trees and crops reproduce themselves, and it's pretty likely that we will continue to produce waste.

Most biomass plants today rely primarily on industrial waste from places like saw mills and the like since they produce a steady supply of uniform material that's easy to burn. However, there is another way to use biomass to produce energy that has attracted a lot of attention lately, thanks in part to some cute branding that refers to one method in particular: cow power.

Whether collected from decomposing manure or just rotting trash in a landfill, methane can be captured and burned to generate electricity. Methane capture is becoming more cost-competitive, and also carries a compelling added benefit in the avoided release of all that gas into the atmosphere. Methane is twenty-five times more potent as a greenhouse gas than CO_2, so diverting it into energy production represents a major improvement.

A Bad Case of Gas: Livestock and Global Warming

There is one hitch with animal-derived methane, and that is that manure from livestock gives off not just methane but carbon dioxide and trace amounts of hydrogen sulfide, which is particularly problematic because it is corrosive—not so good for turbines and other equipment. To be used in generating electricity, methane on the farm has to be purified, which involves using sophisticated *scrubber technology* to eliminate the impurities. Still, the technology is increasingly attractive to farmers since the power they produce can be used on-site to run the farm.

Capturing methane from animal manure is a bit more complicated than simply holding a vacuum hose over a pile of dung. To make the process as fast and, shall we say, as bountiful as possible, the manure has to be stored in a warm (100°F) environment with no oxygen. This type of chamber is called an *anaerobic digester*, and is ideal for bacteria to break down the manure. The process releases a lot of gas, and about 90 percent of it is methane, which after going through the scrubber can be piped directly into a turbine or natural gas engine. The leftover material is still useful, too. The liquids are drained off and used as fertilizer, and the dried out solids can be used in compost or as bedding for animals.

Methane capture is a particularly compelling proposition since it addresses global warming and climate change in two ways simultaneously. When you consider that there are 100 million cattle in this country today,[3] the potential is staggering. And that doesn't even count the pigs and chickens.

On the other end of the cow, however, things are not so rosy. Cows produce an enormous amount of methane by burping. One animal puts out about 280 liters a day and U.S. cattle in total produce 6 million metric tons, again per day. No one has yet figured out a good way to capture cow burps, so this might be one emission source we just have to live with.

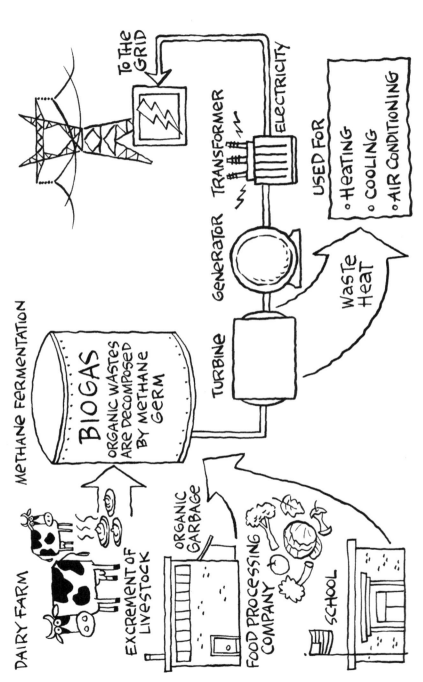

Figure 4.1 Diagram of a biomass processing plant. (Doug Jones)

Geothermal Energy 5

Geothermal energy has been around since the earth's formation and has been used by people for millennia. The word literally means "earth-heat," which is a fitting description since the heat behind geothermal energy comes from deep inside our planet.

You may recall from your elementary school science classes that the rock, soil, and water we know as "earth" are really just our planet's skin, the crust. The continents float on a thick layer of molten rock called *magma*, and if you start digging into the earth's crust, the closer you get to the magma, the warmer it gets in the hole.

In fact, the temperature of the surrounding rock increases about 3°C for every 100 meters, or 5.4°F for every 328 feet.[1] If that sounds like a lot, well it is. Ask anyone who's ever worked in an underground mine. Even if it's well below freezing at the surface, the work environment in a 1,500 meter (4,920 feet) mine can be sweltering. At 3,000 meters (9,850 feet) the rock is hot enough to boil water.

Underground water sources sometimes come into contact with such hot rock, heating the water even beyond the boiling point. (It doesn't turn into steam until it's exposed to air.) If this super-heated water finds its way to the surface via cracks in the rock, you get a hot spring or a geyser depending on the particular characteristics of the rock formation.

Tapping the Earth's Heat: Geothermal Heat and Power

And now, class, what do we know about heat and the production of useful energy? If you've read this far, you probably already have a good idea of how geothermal power works. But long before anyone ever dreamed of

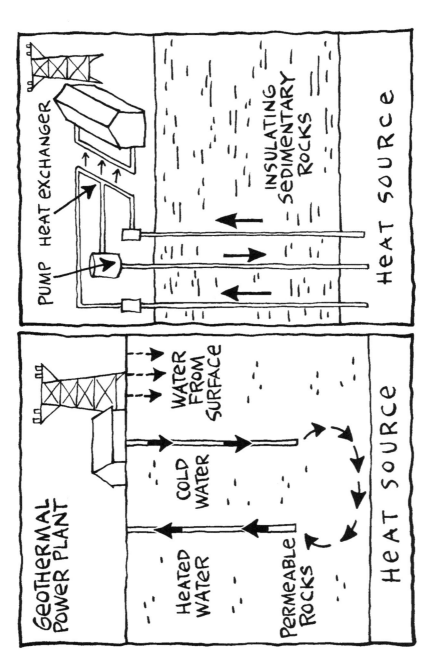

Figure 5.1 Two types of geothermal power plants. (Doug Jones)

generating electricity from the earth's hot springs, the water itself was used for bathing, cooking, and generally feeling better.

Even today, geothermal hot water is used to heat buildings, greenhouses and entire city blocks to say nothing of spas.

Underground steam can also be used to generate electricity. A geothermal power plant uses the same steam turbine technology as a coal-fired plant, but the steam shows up ready-made by the earth via pipes placed into the ground. The white "smoke" that is typically seen rising from geothermal power plants is actually steam given off in the cooling process. The cooled water can then be pumped back below ground to be reheated.

Fun Fact: "Steamy" Iceland

Figure 5.2 The Krafla geothermal power plant in Iceland.
(Ásgeir Eggertsson/Wikimedia Commons)

Our vote for the country with the most ironic name goes to Iceland. Its location in the frigid North Atlantic and its generally cold weather contrast sharply with the island nation's vast geothermal resources. Iceland is home to at least twenty-five active volcanoes and all that nearby lava makes for a great number of geysers and hot springs. Geothermal power plants generate about one quarter of Iceland's electricity, and a whopping 87 percent of the country's buildings use hot water derived from geothermal heat.[2]

Why Aren't There More Geothermal Power Plants Around the World?

Geothermal energy today accounts for less than half a percent of total energy consumption around the world according to the International Energy Agency.[3, 4] Given the obvious benefits of geothermal, why don't we get more of our energy from this source?

In a word, geology. Iceland and certain other select areas (California has fourteen geothermal plants) are blessed with an abundance of geothermal resources. Unfortunately, the rest of the earth is not so lucky. Geothermal formations occur where the earth's crust is thin, and where there is an accompanying supply of underground water. These formations can be rather delicate, however, and in fact some geothermal plants have been shut down due to earthquakes caused by the plant's impact on pressure in the underground reservoirs.

Drill, Baby, Drill

Google's mission is to organize the world's information and make it universally accessible and useful. To do this, they employ extraordinarily sophisticated computer programs that can search, sort, index, and drill through mountains of raw data.

In 2008, Google took its "drilling" in a completely different direction, announcing that Google.org will invest nearly $11 million in technology to expand the nation's geothermal reserves.[5] That's more than the entire U.S. government spends on geothermal projects in a given year!

The word *drilling* in the context of energy usually alludes to oil, but it has a role in geothermal power too.

Geothermal power plants as we know them today are situated where the earth's crust is thin and hot water and steam make their way to the surface easily. But the next generation of geothermal plants, called *Enhanced Geothermal Systems* (EGS), takes a more aggressive approach.[6, 7] Rather than having the steam come to the power plant via natural forces, it is created by pumping water deep underground where it comes into contact with very hot rock and returns to the surface as steam.

Scientists at MIT have estimated that an investment of $1 billion per year over 40 years could yield 100 gigawatts of new geothermal power.[8] That's pretty compelling when you consider the oil and gas industry has already developed highly advanced drilling technology. But while the basic technologies all exist, no one has yet stitched them all together in a demonstration project.

Bottle That Electron!
(Energy Storage)

6

THE ENERGIZER BUNNY IS ONE of the most recognizable pop culture symbols for endurance, but it actually got its start as a parody of a TV commercial for rival battery maker Duracell. Even today, Duracell bunnies can be seen on European and Australian TV, but the Energizer bunny reigns supreme in the U.S.

We've covered the renewable energy landscape and if there is one thing that comes up again and again as a stumbling block for these resources, it's the fact that the energy they supply is variable. Photovoltaic cells do not produce electricity at night, wind farms do not generate electricity on calm days, and ocean waves have to be strong enough for them to be tapped into for electricity generation. The fact that renewables don't make up a greater portion of our energy supply can be attributed at least in part to the lack of large-scale, economical energy storage.

Let's take wind energy for example. For the most part, wind farms in the U.S. are attached to power grids large enough to accommodate their fluctuations in output. This means that the local utility has other types of power plants ready to smooth out the variations in power created by a wind farm. Usually these are natural gas–fired plants that can ramp up quickly, within minutes. There have been incidents, though, where a lack of wind forced some electric utilities to cut power to big industrial customers.

Energy supply from renewable sources can be spotty but our energy consumption needs on the other hand are 24-7. In fact, some proponents of nuclear power and advanced coal plants point to this particular deficiency with renewable energy as its Achilles' heel. They insist that renewable energy will forever be a marginal provider of energy until we are able to generate a predictable stream of electricity to power our homes, offices, townships, and cities during the day and night, during windy times as well as dry and balmy times.

Storing energy in large quantities is the key to making wind and solar energy more useful and for mainstream adoption of renewable energy sources generally. Much like the Energizer bunny that keeps our toys, flashlights, and calculators running, giant batteries offer one way to tap into renewable energy sources in a more predictable way. As we'll see, though, there are other energy storage technologies out there that have nothing to do with drum-beating rabbits.

Most all energy storage devices convert electricity into some other form of energy and quickly convert it back to electricity when needed. The energy is stored as chemical energy, mechanical energy, thermal energy (heat), or even light (as radiant energy).

Some storage technologies are very mature while others are not quite ready for prime time. But our overview will give you a sense for what our world could look like a few years from today.[1] Picture a whole slew of these storage devices sprinkled throughout our power grid, much like the tiny plastic homes and hotels sprinkled around the busy board layout of a game of Monopoly.

Energy Storage Using Water: Pumped Storage Hydro

Pumped storage is a type of hydroelectric power plant, and is presently the most cost-effective way to store large amounts of energy.

Pumped storage power plants have two lakes, one at a higher elevation than the other. The two lakes are connected by a tube called a *penstock* (picture a giant concrete drinking straw). A turbine is placed at the bottom end of the hollow tube. When the water is released from the upper lake, it rushes down through the tube towards the lower lake, thereby turning the turbine. This water has "moving energy," also called kinetic energy. The water collected in the upper lake has "stored energy" called potential energy. The spinning turbines, in turn, drive the generators that produce electricity in the same way as any hydropower plant. The only difference is what happens to the water after it passes through the turbines.

During periods of peak demand for power, water is allowed to drain out of the upper lake into the lower lake thereby generating electricity.

Later, the water from the lower lake is pumped back to the upper lake using a giant pump. The pump motors are powered by electricity from the power grid and this process usually takes place overnight when the energy demand is not so high and electricity prices are lower.

There are approximately 30 to 35 pumped storage power plants in the U.S. today. They are extremely effective and some of them can achieve maximum energy output within 16 seconds of starting up. As noted earlier, all hydropower plants are also very good at following the variations in demand over the course of the day, so they are also well suited to evening out the variations in output from wind farms.

Energy Storage Using Air: CAES

Compressed air can be used as an energy storage device. These devices are known in the industry by the acronym CAES, which stands for *compressed air energy storage*, and they work similarly to pumped storage hydro. During off-peak hours, electric power from the grid is used to compress air and pump it into an underground storage area, such as a cavern or an unused limestone or coal mine. When it is time to release the energy, the air that is trapped in the cavern is used to feed a gas-fired turbine generator which in turn generates electricity.

Another approach uses wind turbines (instead of electricity from the grid) to run an air compressor inside a giant storage tank. The compressed air is released from the tank to generate electricity when needed by spinning a gas-fired turbine.

Because a compressed air–fed turbine can power up faster than conventional power plants, it is ideally suited for use during times of very high energy demand. Compressed air storage sites have been up and running in Germany since 1978 and in Alabama since 1991.

Fun Fact: Did You Know?

Many dentists use equipment that is similar in concept to compressed air storage devices. The next time you are at your dentist's office for a regular checkup, notice the many suction devices and drills on the chair. They are called pneumatic devices and are driven by compressed air as opposed to being run by electricity from the grid.

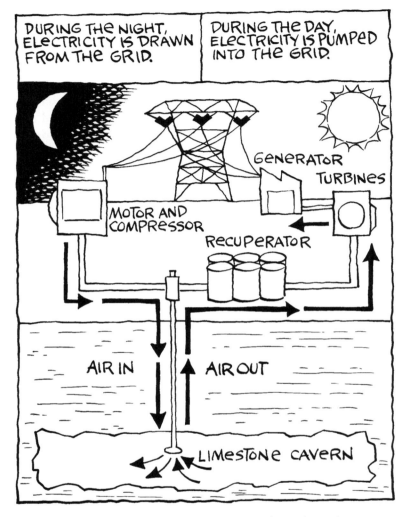

Figure 6.1 CAES systems store compressed air to be used
to generate electricity later. (Doug Jones)

Energy Storage Using Chemical Energy: Batteries

Batteries provide power to a lot of equipment we use in our daily lives. From our automobiles to cell phones to laptops and flashlights, they are fairly ubiquitous. But have you heard of batteries that can power an entire city? (It's true!) How about one that is the size of a double-decker bus?

Giant-sized batteries supplying electricity to the electricity grid itself are a reality today.

All batteries use some kind of chemical reaction to produce electricity, so they all work on the same principle as Volta's "pile" of metal and acid-soaked paper did over two hundred years ago. However, the current only starts to flow if something connects the positive and negative terminals of the battery, like wires leading to your car's starter, or the metal innards of a flashlight.

Different types of batteries use different combinations of chemicals to power their reactions and produce electricity, each having their pluses and minuses (pun intended). Zinc-carbon batteries are what we all know as a standard battery, and they are used in most inexpensive AAA, AA, C, and D dry-cell batteries. Zinc and manganese oxide are used in common alkaline batteries like Duracell and Energizer. Lithium-ion batteries are the type found in hybrid cars and laptops.

But when it comes to storing energy in massive quantities (such as powering an entire township), sodium-sulfur batteries seem to be a pretty good bet. Also called NaS batteries (for the chemical symbols for sodium and sulfur), they consist of liquid (molten) sulfur at the positive electrode and liquid (molten) sodium at the negative electrode. They use a durable porcelain-like material to bridge their electrodes, and generally speaking have a life span of about fifteen years. The batteries are typically charged from the grid at night, when energy demand and prices are low, and discharged during the day when power usage peaks, or in emergency situations.

Japan has made substantial use of sodium-sulfur batteries in grid applications, enough to power 150,000 homes. American Electric Power, a major utility servicing eleven Midwestern states in the U.S., will soon have installed a total of 10 megawatts' worth of energy storage devices throughout their service territory.

On a dollars-per-watt basis, though, sodium-sulfur batteries cost (at $3/watt) about the same as building a new power plant. However, the technology is promising—sodium sulfur batteries are compact, long lasting and efficient, but they still have a long ways to go before the economics become viable.

Energy Storage Using Mechanical Energy: Flywheels

A flywheel refers to a large cylinder or wheel that is brought up to speed by a generator during the night. During the daytime, it runs a turbine, which generates electricity until the wheel decelerates.

The flywheel system works on kinetic energy, which is the motion of the wheel. Think of it as a mechanical battery, spinning at very high speeds (>20,000 rpm) to store energy that is instantly available when needed. They are best suited for standby power applications. Flywheels are not built for long-term energy storage ("long-term" in this business being twelve hours), but they're fine for evening out minute-to-minute variations in supply and demand on a grid. Despite their massive size, flywheels are very reliable and can run for years with almost no maintenance.

Superconducting Magnetic Energy Storage (SMES)

Now we're getting into the Flash Gordon area of energy storage. SMES devices store energy in magnetic fields created by the flow of current in a superconducting coil that has been super-cooled. They are capable of delivering a large amount of power almost instantaneously, a major advantage over pumped hydro or CAES where there is a time lag while the water or air makes its way to the turbine. The only drawback with SMES devices is that the power they deliver is short-lived and thus only suitable for momentary adjustments in electricity supply.

Supercapacitors and Time Travel

There are far too many other emerging technologies to cover them all in this chapter. They include electrostatic capacitors that store energy as electric charge; electrochemical capacitors that can be charged and discharged hundreds of thousands of times in comparison to batteries, which can only withstand thousands of discharge cycles; supercapacitors that offer very high capacitance in a small package; the list goes on.

Whether the winning technology will be based on an array of electrostatic nanocapacitors or plutonium-fueled portable flux capacitors (that's fiction—remember Doc Brown's time machine in *Back to the Future*?), all of them share common design goals such as high energy density, high power output, fast recharge and discharge capability, and of course the clincher, affordable cost.

Conclusion

According to the Energy Information Administration, the share of U.S. generation coming from renewable energy sources (including hydropower and biomass) will grow from 9 percent in 2008 to over 17 percent in 2035. Wind and solar energy contributed a meager 2 to 3 percent of total U.S. generation. From this modest starting point, there is no doubt that we can add at least an order of magnitude more renewable energy sources to our mix before energy storage becomes a serious constraint. But to entertain ambitions of dethroning "King Coal" (the source of half of the electricity consumed in the U.S), renewable energy sources have to be coupled with economically and technologically feasible energy storage devices.

Dependability is a key issue for all energy sources. Coal, natural gas, and nuclear power plants have it. Renewable energy sources have less of it. If we could bank energy when it is abundant and inexpensive and release it at other times when needed, we would have a far more reliable and economical energy grid. It would be more environmentally sound too, as energy storage seems to be the linchpin of renewable energy's success.

Power outages are responsible for $80 billion in economic losses per year, according to studies done at the Lawrence Berkeley National Laboratory. Many of them are caused by trees falling on transmission lines, power surges from lightning strikes or unpredictable shifts in supply hurting the grid's ability to meet energy demand instantaneously. When supply does not match demand by a wide enough margin, you have the makings of an outage.

Using the right combination of energy storage technologies and distributed generation, utilities could also defer or possibly even avoid construction of new transmission lines, substations, and power plants.

Energy storage devices sprinkled liberally throughout our power grid could ease pain on all these fronts, facilitating mainstream adoption of renewable energy sources and making our grid more reliable by greatly reducing the number and severity of small outages. Gas-fired power plants make up the vast majority of generation used to serve peak demand, so there's an eco-friendly benefit attached to many energy storage technologies if they can replace those peaking units. However, only time will tell which technology (or technologies) will race ahead of the others to become economically viable for large scale adoption.

Energy and Transportation

THE SCREWBALL COMEDY OF THE 1980S *Planes, Trains, and Automobiles* showed the error-prone adventure of the main characters Steve Martin and John Candy as they work their way back home from New York to Chicago for Thanksgiving after every available mode of transportation fails them.

So what do cancelled flights and traffic jams have to do with a book on energy? In a word, everything.

Transportation is one of the largest users of energy and is also something of a barometer for energy costs in general. When the news reports shout "energy prices up," they're usually talking about gasoline, or more generally, oil.

Of course transportation is not limited to automobiles and the energy that underlies it is not limited to petroleum. For the purposes of this book, though, we've elected to focus on cars for the simple reason that they represent the primary mode of transport for most Americans and they are also a flashpoint for a wide range of debate that connects energy to issues like land use, public safety, and the economy.

In this part, we'll explore the complex symbiotic relationship between energy and cars. Buckle up—this ride could get a little bumpy.

Land, Sea, and Air— Energy in Transportation

7

HAVE YOU EVER HEARD THE EXPRESSION, "drinking from a fire hose?" Well, that's kind of what its like learning about energy and transportation. We could write a book—several books—just on this area of the energy world. Others already have, so rather than try to duplicate what they've done, we thought we'd focus on personal travel and people's love affair with the automobile.

However, any discussion about cars should be placed in some sort of context amid the other modes of transportation: air, sea, and rail. (Interplanetary and metaphysical modes of transport will not be covered.) So, what follows is a very brief stopover in the mass transit/cargo sector before we move on to a more detailed look at energy in automobiles.

The First Vehicle

OK, to be precise, human beings' first mode of transport—one still popular in many locations today—was walking. After a while we figured out how to get animals to do it for us, and that worked pretty well. But the first big breakthrough in long-distance travel was the boat, and in particular the sailboat. To people with the ability to traverse thousands of miles in a matter of weeks, the world would never be quite as large again.

When it comes to land transportation, you have to go back to 3500 BC for the earliest record of what might have been a wheeled vehicle, an image of a wagon with four wheels shown on a clay pot discovered in Poland. But humans were plying their local rivers, lakes, and oceans in small watercraft long before that.

Energy Use in Ships

There are over 55,000 merchant ships in operation around the world to-day, and collectively they handle over 90 percent of world trade, transport-ing every kind of cargo including energy (crude oil, liquefied natural gas). Most modern ships rely on some form of diesel fuel to turn their propellers. Diesel emerged as the preferred engine type for marine use in part because the engines don't have spark plugs, which could be adversely affected by the moisture and salt in the air. You may have heard the phrase "bunker fuel" and this is a blanket term used to describe all kinds of marine fuel, whether marine diesel or the heavier #2 diesel. The term comes from the days of coal-powered ships when that fuel was stored in portside bunkers.

All marine fuel is "dirtier" than gasoline or even the diesel we use in our cars—there are different regulations governing the emissions from seagoing vessels. Some ships even use heavy fuel oil that must be heated with steam in order to flow. In some areas (e.g., Alaska), ships are required to burn fuel with low sulfur content in order to cut down on emissions. In recent years, ships have begun to make use of gas oil turbines and even electric propulsion systems as an alternative to diesel.

Fuel efficiency is very much a moving target when it comes to ships. There are many variables, and in the end, trying to compare ships and cars is a bit like comparing the proverbial apples and oranges. For starters, ships consume a lot of fuel, so rather than miles per gallon, we're talking more about gallons per mile. Oh, and did we mention a mile isn't really a mile at sea? A nautical mile is equivalent to 1.15 land miles because a nautical mile is actually a fraction of the circumference of the earth at the equator (1/21,600th to be exact).[1] Speed, then, is not measured in land miles per hour, but in nautical miles per hour or knots.

OK, so if we want to get even just a rough idea of fuel consumption in ships versus those in cars, what else do we need to keep in mind?

Perhaps the most important variable is *carrying capacity*. Ships use a lot of fuel, but they can carry an enormous amount of weight, whether it's cargo or passengers. Most fuel efficiency numbers for cars are calculated with two passengers in mind, but in the case of ships this capacity can vary widely thereby impacting the overall fuel efficiency. In fact, the *Queen Elizabeth II* has been estimated to have gotten about 18 miles per gallon per passenger at cruising speed.[2] (The ship could carry 1,900 people and required a gal-lon of fuel to go 50 feet, or 0.00947 miles. Then you have to adjust for imperial gallons, and well . . . just trust us on this one.)

Eighteen miles per gallon isn't exactly outstanding in car terms, but we're betting it's higher than you would have thought for what was until relatively recently the largest passenger ship in the world. Some newer cruise ships utilize diesel engines to generate electricity which in turn is used to operate large electric motors that actually do the work of turning the propeller. This arrangement has a few advantages in addition to better fuel efficiency— which can be as much as 20 percent over conventional drive systems—including more room for passengers and cargo, and less vibration and noise.

(Full disclosure: Bob's employer, ABB, manufactures these systems and while he's never set foot on a cruise ship, he still thinks electric marine propulsion is pretty cool.)

Rail Travel

It's hard not to love trains, unless of course you happen to live near an active railway. We once shared an office situated a matter of yards from a busy railroad line in Oakland, California. The building was converted from an old cannery, and the only thing separating us office workers from the rumble of the freights going by was a thin wall of cinderblock and single-pane windows. Not especially good for conference calls.

Rail travel, however, has had an enormous impact on transportation, and is poised for resurgence in our carbon-constrained future. But to start at the beginning, you have to think steam. Of course, the steam engine had been around for years before it was coupled to a railcar, but it wasn't until 1797 that the first steam locomotive was built. Whether fired by coal or wood, the steam created would drive pistons in a reciprocating engine like the kind you find in automobiles except that the pistons move from the pressure of the steam rather than the pressure of expanding gases during combustion.

The steam engine reigned supreme all the way to the twentieth century when steamers began to be replaced by diesel and electric locomotives, which still account for the vast majority of passenger trains today.

To be clear, when we say "diesel" in the context of modern trains, we're really talking about electric motors being powered by a diesel engine that generates electricity. "Electric" locomotives simply draw power from an overhead line or third rail rather than generating it onboard.

There are certain advantages and disadvantages to both of these designs. Diesel locomotives are cheaper to buy, and are still relatively fuel efficient. According to the American Association of Railroads, a typical diesel train can move a ton of freight over 400 miles on one gallon of fuel. Diesels are also not susceptible to power grid disruptions since they make their own electricity.

All-electric locomotives, on the other hand, are lighter in weight (no need for big diesel engines or fuel tanks), which allows them to travel at higher speeds. The current records for electric and diesel engines are telling in this regard: 320 mph for the electric and 148 mph for the diesel. Electric locomotives also benefit from having all the power of the grid at their disposal versus the limited supply of a diesel engine, which also contains a lot of moving parts and so requires a level of maintenance that electric trains don't. Electrics are also quieter and produce lower energy losses in the form of heat, so more power gets to the wheels.

Electrics seem to be the better pick, especially for passenger travel, so why don't we see more of them? As you might have guessed, it often comes down to cost. The electric power supply needed to run all-electric trains requires more infrastructure than simply building the tracks. This might make sense in urban areas where power is easily accessible, but on long haul routes the up-front investment can be prohibitive.

Trains using magnetic levitation ("maglev") were conceived more than a century ago, but today there is only one in commercial operation, in China.[3, 4] The idea is simple, really. One set of magnets on the train pushes away from another set of magnets on the track, allowing the train to levitate a few inches. Another system of magnets along the railway uses magnetic force in a push/pull fashion to move the train forward. The elimination of friction and the reduced weight of these trains make them highly energy efficient, but—here we go again—they are very expensive to build.

Hybrid drive systems have also begun to appear in trains. Some "switching engines"—locomotives used to move cars around in a rail yard—have multiple engines on board that cycle on and off as needed. Others use a small diesel engine to charge batteries that in turn power electric motors. If this setup sounds familiar, it's probably because it is the same concept that Chevrolet is using in its forthcoming hybrid car, the Volt.

In fact, one of the key technologies behind hybrid cars owes its development to railroads. Regenerative braking captures the energy otherwise lost as heat through brake pads and rotors while bringing the vehicle to a stop. Trains have used this technique to recycle energy for many years, but it's only been with the introduction of hybrid cars that the idea of "regen" entered into our automotive consciousness.

The Efficiency Question: Apples to Apples?

To say that it's difficult to compare various modes of transportation, or even different vehicles of the same mode, would be an understatement.

There are so many factors that go into determining the actual efficiency of a given vehicle operating under certain conditions, it's almost impossible to arrive at a true apples-to-apples comparison. Trains are especially difficult to compare as their operating conditions, fuel type, and routes all vary widely.

That said, it's worth looking at a variety of reports to get a sense of the energy use question at least in broad strokes. Table 7.1 is derived from the U.S. Department of Energy's (DoE) annual *Transportation Energy Data Book* for 2008. The figures for each vehicle type represent an average of the various cars, buses, trains, etc. currently used across the country.[5]

The "vehicle MPG" is simply the fuel efficiency of the vehicle itself; "passenger MPG" accounts for the energy used *per person* in the vehicle. When you consider how many people the given vehicle is moving, some interesting things might happen to your impressions (vanpools are number one?! Really?). As the table shows, trains are terribly inefficient on their

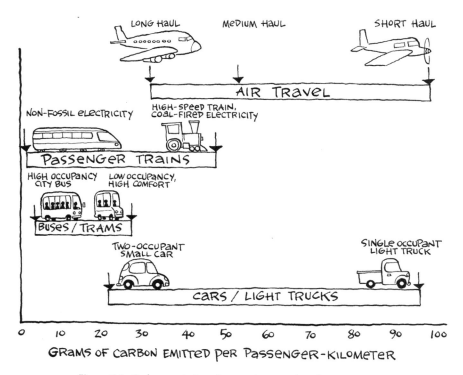

Figure 7.1 Carbon emissions from various modes of transportation.
(Aviation Environment Federation [UK] and American Bus Association)

Table 7.1. Fuel efficiency in miles per gallon (or equivalent) of different vehicles

Mode	Vehicle MPG	Passenger MPG
Vanpool	14.2	87.0
Motorcycles	51.7	62.0
Rail	1.7	40.8
Cars	20.9	32.7
Personal trucks	17.0	29.2
Buses	3.1	27.1

Source: U.S. Department of Energy.

own, but when you factor in the hundreds of passengers they carry their actual passenger-miles per gallon of gasoline equivalent comes in at a respectable third.

Again, DoE's numbers represent an average, so if the U.S. were to implement more high-efficiency locomotives, more all-electric trains or maglev trains, the mpg equivalent for rail travel as a whole would increase. To illustrate how much difference there can be between averages and specific cases, and also between studies, check out figure 7.4 later in this chapter; it breaks down the various vehicle types into specific technologies and models.

A Rail Travel Renaissance?

Given the advance of technology and its impact on efficiency, rail travel could be on the brink of a major revival in the U.S., both in urban (e.g., subways, light rail) and long-distance (e.g., high-speed rail) transport. According to the Association of American Railroads, America's freight rail system has improved its fuel efficiency by a whopping 85 percent since 1980. The graph below paints a compelling picture—miles traveled by our nation's freight trains have nearly doubled but the amount of fuel used has barely risen at all.[6]

Whether improvements of this magnitude are possible in passenger rail remains to be seen, but even the Department of Energy's snapshot of our current transportation options is reason for optimism. What's more likely to hold up growth in the U.S. passenger rail system (or as some might say, the implementation of a "real" passenger rail system) is the difficulty in building new rail lines, especially in more densely populated areas. If you think NIMBYism is bad for power lines, try siting a train track. Can you say court challenge?

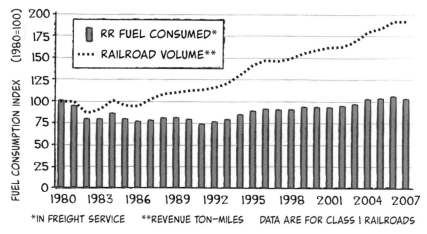

Figure 7.2 Railroads today move almost twice as much freight using nearly the same amount of fuel as they did thirty years ago. (Association of American Railroads)

Transportation Takes Off: Air Travel

Many people know that Orville and Wilbur Wright invented a motor-driven flying machine that was the precursor to today's airplanes. Some may also know that the Wrights' day job was fixing bicycles, but we're willing to bet most don't know that the bicycle was, at that time, still a relatively new invention itself. The modern form (with pedals, chain, and rear-wheel drive) appeared less than twenty years before the first flight at Kitty Hawk in 1903.

By contrast, it would be decades after its introduction that air travel would become the mass people-mover it is today. In the interest of brevity, we'll skip directly to the jet-powered aircraft that most of us are familiar with. Those planes run on jet fuel, which began as kerosene or a gasoline-kerosene mixture and is much lighter than the heavy diesel used in ships. Jet fuel also contains additives to prevent sparking from static electricity and to inhibit corrosion of engine parts. Jet fuel typically makes up about 10 percent of the output from one barrel of crude oil, and as one of the lighter fuels its yield can be boosted slightly in the cracking process[7] (for more on cracking and how to get more from a barrel of crude than you put in, see Volume 1).

The cost of fuel is second only to labor in the airline industry, and the run-up in oil prices in 2008 produced a similar spike in jet fuel, as shown in figure 7.3. And, just like drivers in one region pay significantly more for gas than drivers in other regions, so too do airlines pay varying prices

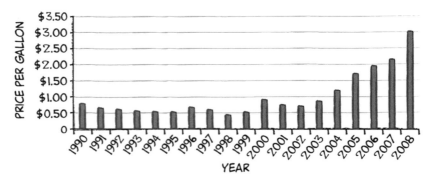

Figure 7.3 Jet fuel prices have hit the stratosphere in recent years. (EIA)

for jet fuel in different locations. The reasons are similar: limited refining capacity, limited storage capacity, and constrained distribution thanks to high mountains. All of these factors make jet fuel on the West Coast more expensive than the same product sold east of the Rockies.

The U.S. is a net importer of jet fuel, using 1.63 million barrels per day while producing 1.44 million barrels. Airlines are understandably interested in reducing fuel consumption, in part because there is no viable substitute for jet fuel. The industry is also coming under increasing pressure to mitigate its CO_2 emissions.

Air travel, according to the industry, accounts for about 2 percent of global CO_2 emissions, but the number of flights and passenger miles traveled is increasing every year. In fact, the Federal Aviation Administration predicts that passenger volume in the U.S. is set to double over the next ten years, and could even triple by 2025. Meanwhile, more and more people around the world are moving up the economic ladder to a place where air travel is within their reach. All of that spells trouble for an industry wedded to a fuel that produces three pounds of carbon in the atmosphere for every pound of fuel burned.

Energy Deathmatch: Boeing 747 vs. Toyota Prius

Let's get ready to rumble. In this corner, from Yokohama, Japan, with a wheelbase of 106 inches and weighing in at 2,920 lbs, the world's most popular hybrid car, the Toyota Prius. And in this corner, from Seattle, Washington, with a wingspan of 200 feet and weighing in at 393,000 pounds, the original jumbo jet, the Boeing 747.

Are you kidding? This is David and Goliath. This is U.S. vs. USSR in hockey at the 1980 Olympics. The smart money is on the little guy, right? Not so fast—the real story is more complicated.

According to aircraft manufacturer Boeing, the company's flagship 747 uses about 5 gallons of fuel per mile. On a ten-hour flight, the plane might burn 36,000 gallons of fuel, which sounds almost unfathomable. (How does it even get off the ground with that much weight?) The Toyota Prius gets 46 miles per gallon (combined city/highway driving) according to EPA estimates.[8]

But when you look at fuel economy *per passenger*, the 747 does pretty well. Even at only three quarters full (i.e., 300 passengers), the 747 is getting roughly 60 miles per gallon per person at cruising altitude, and around 48 mpg when you factor in taxiing, takeoff and landing. It's also moving at a top speed of 550 miles per hour. Not bad when you compare it to the average car, which gets around 25 to 35 miles per gallon at 65 mph. Granted, most people drive faster than the speed limit, but that only makes the car less efficient.

Now, the Prius is the most fuel-efficient production car on the road in the U.S. (pipe down Tesla fans—we'll get to your dreamy roadster in a minute). But even with a highly speed-conscious driver, the Prius is still only about even with the big jet in fuel economy on a per-passenger basis.

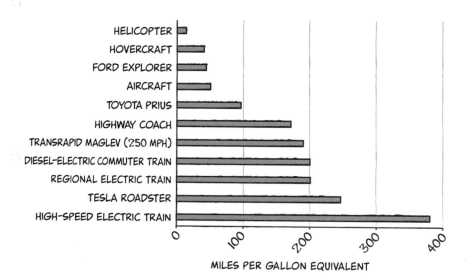

Figure 7.4 Passenger miles per gallon (equivalent) across various types of transportation. (James Strickland)

Of course, there's a bit more to the fly vs. drive question.[9] First, there are a lot of factors that impact the fuel efficiency of both aircraft and motor vehicles. Planes have a wide range of emissions and fuel economy, they fly a wide range of distances, and they do so with varying amounts of passenger and cargo weight. Cars are no less variable—just compare the Ford Focus to the Ford Excursion—and they are driven under similarly various circumstances.

In terms of CO_2 emissions, the playing field is similarly large. CO_2 emissions for jet planes run from about 32 grams per passenger-kilometer to around 100 grams. Cars fall into a range of approximately 20 to 100 grams per passenger-kilometer. However, planes get an extra strike against them because they release their CO_2 high up in the atmosphere, which in turn has an impact on how strongly the carbon affects radiation entering the earth's atmosphere. This is known in the climate science world as "radiative forcing" and the Intergovernmental Panel on Climate Change has come up with a multiplier of 2.7 to account for air travel's increased climate impact as a result of it. So, a gram of CO_2 released at high altitude has about 2.7 times as great an impact as a gram of CO_2 released at sea level.

So, to more accurately compare the impact of emissions from traveling by plane vs. car, bump up the air travel option's CO_2 emissions by a factor of 2.7 and then compare it to the driving option. Bottom line: planes emit less CO_2 than cars, but what they do emit has a much greater impact on the sun's radiation and in turn on climate. On the other hand, if you're just interested in fuel economy, you might want to reconsider that four-hour drive vs. one-hour flight.

Fun Fact: Carbon Calculators for Air Travel

Chooseclimate.org has a carbon calculator that accounts for takeoff, cruising, and landing to come up with a total emissions figure for any given flight. Full details are provided of the method of calculation: http://www.chooseclimate.org/(click on the "Flying off to a warmer climate" link).

Fuel Costs for Airlines: How Do They Cope?

Airlines. Except maybe for the phone, cable, or utility company, is there another industry that is capable of inspiring the same level of seething hatred among its customers? They overbook your flight, they lose your luggage, and on top of everything they have the nerve to charge you extra if you want to check a bag. Before you know it, we might have to drop a couple of quarters into a slot before using the toilet (you heard it here first).

Flying today provides a very clear, if unpleasant, example of how closely energy and transportation are linked. Every time jet fuel prices go up by a penny per gallon, it costs a typical airline an extra $25 to $30 *million* per year because collectively the airlines use 2.5 to 4.0 *billion* gallons in a typical twelve-month period. A jump in fuel prices by 5 or 10 cents per gallon can financially cripple an entire company.

The price of jet fuel in the U.S. rose by 115 percent between 2000 and 2007. By the spring of 2008, crude oil was scratching at the door of $150 a barrel and airlines were looking at massive cost containment efforts, rising fares and a very uncertain future.

So how do airlines cope with this?

Progressive airline companies have resorted to sophisticated financial techniques called *fuel hedging* to manage their unpredictable fuel costs.[10] Hedge contracts are not that different from the auto insurance policy you have with your insurance company. You pay the insurance company a premium, generally every six months or so, and in return they will provide a specific amount of coverage for your car.

In a similar way, hedging contracts allow airline companies to buy "insurance" against spikes in fuel prices. But the hedge contract requires an upfront cash payment (i.e., the "insurance premium"). Provided you can come up with the money, fuel hedging is an effective strategy and is employed not only by airlines but by electric utilities and trucking companies as well.

There was one airline that played its hedging hand superbly during the recent boom in oil prices: Southwest Airlines.[11] The company locked in fuel at a crude oil equivalent of $70 a barrel. As prices soared to double that amount, Southwest looked to be in the catbird seat. Ah, but how quickly things change.

The financial crisis infecting the world's credit markets soon spread to the wider economy and oil prices crashed. By the end of 2008, Southwest was still holding fuel contracts at $70 a barrel, but the market price had dropped so that the company was now paying 35 percent above market rates. There's probably some kind of metaphor for this that combines Andy Warhol's fifteen minutes of fame with the "what comes around" adage, but you get the idea.

There is a second way for transportation companies to insulate themselves from fuel price fluctuations, and that is simply to invest in fuel efficiency technology. Boeing is doing just that with its new 787 Dreamliner. The mid-sized plane makes use of lightweight composite materials and more efficient onboard systems to improve its fuel economy by 20 percent over the company's 767 model.

Fuels from Fossils (Revisited)　　8

O K, NOW THAT WE'VE CRUISED through planes, trains, and ships, it's time to settle in for a more in-depth look at the mode of transportation that is so dear to our modern lifestyle: the automobile. In this chapter we'll focus on fuels derived from petroleum, including gasoline and diesel as well as some lesser known varieties. If you're interested in alternatives like biofuels, head over to chapters 9 and 10 because this chapter is all about driving on petro-power.

A Brief History of Fuels for Automobiles
Humans have tinkered with various forms of energy to fuel transportation. From steam-powered boilers to portable windmills bolted to four wheels, early inventors attempted to tap into many different energy sources to drive wheeled vehicles.

The Italians are known for producing vehicles that will compel middle-aged men to make questionable decisions with their retirement savings, but they don't always deliver the highest level of cargo room or reliability. They got started early. The first to put his ideas on paper was Guido da Vigevano in 1335. His design featured a windmill type drive bolted to gears and wheels. Leonardo da Vinci designed a tricycle driven by a clockwork mechanism. Neither of these designs was actually made into a vehicle, but the Italian tradition in design was under way.

During the 1670s, a Catholic priest named Father Ferdinand Verbiest, while a member of a Jesuit mission in China, is said to have built a steam-powered vehicle for the Chinese emperor Chien Lung.[1] In the eighteenth century, various inventors built stationary steam engines but it was not until 1765 that James Watt successfully built the first pressurized steam engine.

Vehicles powered by internal combustion engines got their start in 1862 when an inventor named Alphonse Beau de Rochas developed the precursor to the engines we know today. His process of bringing a flammable gas into a cylinder, compressing it, combusting the compressed mixture, then exhausting the spent gases came to be known as the Otto cycle, or the four-stroke engine.

Why is it not known as the Rochas cycle, you ask? Well, de Rochas was a talented inventor but not so much of a businessman. He neglected to patent his design, but several years later in 1876, Nikolaus Otto did. De Rochas might have enjoyed bragging rights at the local tavern, but Otto had his meal ticket—or so it seemed.

In fact, years later the work of de Rochas allowed entrepreneurs Daimler and Benz to break the Otto patent by claiming "prior art" from the de Rochas design. In a final twist, Karl Benz was ultimately credited as the inventor of the first "true" automobile (patented in 1886) since Daimler's car was really just a horse carriage with an engine, whereas Benz's had a chassis designed from scratch.[2, 3]

A Fuel in Search of a Fire

While all of this automotive tinkering was going on, the nascent oil industry was busy getting started in places like Pennsylvania, where the first U.S. oil well was drilled, and later in the massive fields of Oklahoma and East Texas. The oil market back then was a roller coaster, swinging wildly from boom to bust. By the early part of the twentieth century, however, there was so much oil being produced not only in the U.S. but globally, the depressed prices threatened to put many an oil man out of business. Then something wonderful (for the oil industry) happened.

Automobiles had been around for decades by this time, but they were largely playthings of the rich. Most people got around on trains, streetcars, and horses. Henry Ford changed all that by introducing a car that was produced in such a way as to make it affordable for the middle class. The ensuing explosion of car ownership in the U.S. remains unmatched in history. The Model T was released in 1908 and by 1918, just ten years later, over 15 million had been sold, accounting for fully half the cars in the United States.[4] From an energy perspective, the arrival of the mass market automobile was equally important because it created a huge market for petroleum fuels virtually overnight. The oil industry was saved!

The symbiotic relationship between oil and transportation drove the U.S. economy and American culture itself through the rest of the twentieth century, and it continues to define much of our national identity today. The American Dream and the automobile are inextricably linked. The car, along with the interstate highway system, made possible the suburban landscape we know today. Whether or not that landscape represents the American Dream made real or a scourge on the face of the earth is a matter of opinion, but it is almost impossible to overstate the impact of car culture on our energy landscape.

One last tidbit before we go, and it may come as a surprise. While Ford's vehicles revolutionized personal travel, Henry Ford himself never envisioned his cars running on petroleum, at least initially. No, our great industrialist was also a major proponent of incorporating nature into industrial processes, and he saw ethanol as the most likely fuel for the automobile. That, of course, is another story that we cover in chapter 10.

Today, close to 84 percent of crude oil is used for producing energy-rich fuels, mostly for transportation, and over 97 percent of automobiles run on petroleum. In the U.S., roughly two-thirds of the oil consumed each year is burned on the nation's roads. Worldwide, motor vehicles number over 750 million, and that figure is expected to triple by 2050.[5] It's worth taking a moment to consider what our energy landscape would be like today had oil and the automobile not married so young.

Fun Fact: Did You Know?

Mercedes-Benz is one of the premier automotive manufacturers in the world, and thanks to its historical link to Karl Benz, it is also the world's oldest continuously produced line of automobiles.

The Butterfly Effect

The tagline for the 2004 movie *The Butterfly Effect* reads, "Change one thing. Change everything." Ashton Kutcher's character suffers from memory blackouts during his childhood and later (painfully) realizes the impact of seemingly small, unrelated events in the past that trigger a series of aftereffects.

In fact, the phrase "the butterfly effect" encapsulates a notion in chaos theory that suggests that small occurrences today can significantly affect the outcomes of seemingly unrelated events tomorrow.

Such is life in the oil industry. The connection seems tenuous at first glance, but by now we're all well acquainted with the effect of this particular winged son-of-a-caterpillar. For every $10 a barrel of crude oil price increase, gasoline prices at the pump rise by about 25 to 30 cents a gallon.

On the demand side of things, oil consumption is what economists call "inelastic." This means that the consumption of fuel is so essential to daily life that increases in prices only curb consumption marginally, if at all. Transportation illustrates this idea well. After all, you are very unlikely to suddenly start biking or riding public transportation to work immediately after hearing the morning news reports of cuts in crude oil production in the Middle East.

At the same time, the fact that most people cannot easily curtail their consumption of gasoline means that even modest fluctuations in oil prices can produce disproportionate variations in prices at the pump. And that will in turn send a shockwave through the entire economy because everything is connected by energy, specifically oil.

Consider that virtually everything you buy is shipped, flown and trucked using that same essential fuel. All of those businesses are committed to using petroleum. Now consider the cumulative effect of rising fuel costs as each maker of widgets must pay more for the sub-widgets he needs, and how each supplier of sub-widgets must pay more for the sub-sub-widgets they need, all because of rising transportation costs. By the time all those layers of additional cost are tallied up in the retail price of a finished widget, the impact of energy on transportation and the larger economy becomes obvious.

Fun Fact: Did You Know?

The Strait of Hormuz is a narrow passageway of water where the Persian Gulf empties into the Arabian Sea. It is considered one of the most, if not the most, strategic strait of water on the planet because through it passes much of the crude oil from Bahrain, Iran, Iraq, Qatar, Saudi Arabia, and the United Arab Emirates to the rest of the world. According to the Energy Information Administration, roughly 20 percent of the global oil supply and 40 percent of all seaborne traded oil (roughly 16 to 17 million barrels of oil) flows through the strait every single day.

Gasoline

One barrel of crude oil, when refined, produces about 20 gallons of finished gasoline (nearly half the total) and 7 gallons of diesel, as well as other petroleum products.

Why is gasoline so dominant as an automotive fuel? For starters (no pun intended, really), there's energy density—a measure of how much energy a given fuel delivers per unit of volume—and gasoline comes in at a fairly high 36.6 kWh/U.S. gallon. What that means is that one gallon of gas packs a pretty big energy punch, 42 percent more energy per gallon than ethanol and slightly less than diesel. Gasoline is also relatively stable—it will keep 60 days easily in proper storage containers—and though it is more volatile than diesel or jet fuel, it is still generally safe to handle.

These characteristics give gasoline an advantage over other liquid fuels, but its place atop the transportation world could only have been made possible by the abundance of oil that in turn made gasoline cheap.

Gasoline is really a mixture of ingredients. It starts with distilling crude oil, but other chemicals (most of which end in "−ene") are added in the refining process. These additives appear in varying quantities depending on where the fuel is produced, and are used mainly to increase the octane level of the end product. Many of them are considered hazardous materials and are tightly regulated. Benzene, for example, is generally limited to 1 percent by volume in gasoline sold in the U.S. MTBE, used as an oxygenate, was originally a replacement for lead but was later found to have some nasty environmental drawbacks of its own with regard to water supplies.

So why go to all the trouble to boost octane? Why not just use "virgin" gasoline distilled straight from oil?

Octane, by definition, is the resistance to burn. The higher the rating, the slower the fuel will burn during the combustion cycle of your engine. If you're scratching your head as to why you'd want the fuel to burn more slowly, bear with us—it really does make sense.

If gasoline burns too fast it will actually detonate in the cylinder, which prevents the fuel from burning completely and can damage engine parts. Having the fuel burn more slowly allows the car's engine to function properly and draw more of the energy out of the fuel. We're talking milliseconds here when we say the fuel burns "slowly," but it makes all the difference. High compression engines require higher-octane fuel because lower-octane gas simply won't hold out long enough under the increased pressure.

Contrary to popular belief though, putting high-octane gasoline in a "regular" car won't improve your fuel efficiency, acceleration or emissions. Ideally, you want to match the correct octane rating of the gasoline to your car's engine design to maximize fuel economy and lower emissions. The only time you may need to switch to gas with a higher-octane level is if your engine knocks.

Fun Fact: Do Race Cars Use the Same Fuel We Do?

NASCAR racers burn 110-octane leaded gasoline. In comparison, the "regular" gasoline that most of us purchase at the gas station is an 87-octane gasoline, and the highest octane available to consumers is still well below 100. Cars racing in the Indianapolis 500 burn pure methanol (a.k.a. wood alcohol).[6] It can run at high compression ratios, meaning it delivers more power even if it contains only about half the energy of gasoline per gallon. The power of the fuel allows the racecars to accelerate quickly, but the lower energy per gallon means fewer miles per gallon of fuel. The decrease in range with methanol is not a problem for racecars though, since all the cars on the track are using exactly the same fuel.

Methanol also has a nice safety feature—you can extinguish a methanol fire with water. Indy cars used to run on gasoline, but moved to methanol following a horrific crash at the 1964 Indianapolis 500 that killed drivers Eddie Sachs and Dave MacDonald. The black smoke from burning gasoline made it impossible for drivers to see what was ahead of them.

Additional Resources on the Web

Want to know where you can find gasoline for the lowest price in your area? Log onto the website for the American Automobile Association (http://www.aaa.com) and click on the AAA Gas Price Finder.

This user-friendly application allows you to type in a home or office address and will locate the gas station offering the lowest price for gasoline within a 3 to 10 mile radius.

GasBuddy.com is another site aimed at helping you find cheap gas prices in your city.

Diesel

Diesel engines were created by German inventor Rudolf Diesel toward the end of the nineteenth century. In 1898, Diesel was granted a U.S. patent for an "internal combustion engine," and the first U.S. production of his design soon followed. His engines were used extensively in a variety of industries outside of transportation such as electric and water power plants, marine craft, mining, and oil fields. The diesel engines of today are all descended from Rudolf Diesel's original design.

What Makes a Diesel a Diesel?

Ok, for those of us who didn't take auto shop in high school (do they even have auto shop anymore?), we need to quickly explain the difference between a diesel engine and the more familiar gasoline variety. Gas engines are also known as spark ignition engines because they use a spark plug to begin combustion. Fuel and air are mixed together in the cylinder, the mixture is compressed, the spark plug ignites it and the resulting force turns a drive shaft.

Diesels, also called compression ignition engines, work differently. They don't mix the fuel and air beforehand, but rather compress the air first, raising its temperature to the point where the fuel will ignite spontaneously when it is injected into the cylinder. The rest of the process works essentially the same way as in gasoline vehicles where the engine turns a driveshaft that goes through a transmission and differential to power the wheels.

Clash of the Fuel Titans: Diesel vs. Gasoline

So which is better, gasoline or diesel? The answer depends on what part of "better" is more important to you. The basic tradeoff between our two great automotive fuel/engine combinations comes down to efficiency, emissions, and operating characteristics.

Generally speaking, diesel engines are more *efficient* than gas, meaning you can go farther on a gallon. This is due to diesel's greater density, both in terms of mass and energy. A gallon of diesel weighs more than a gallon of gasoline, so there's literally more of it in the same amount of space in your tank. There's also slightly more energy in diesel than in an equivalent amount of gasoline.

In terms of emissions, diesels typically produce more particulate matter—the tiny pieces of unburned fuel in a car's tailpipe emissions—than gasoline engines do because the fuel is less refined. So, while diesels get better mileage, they tend to emit more bad stuff. The higher fuel efficiency of the diesel engine results in lower carbon dioxide emissions *per mile* but the higher energy density in the fuel itself results in higher CO_2 emissions *per gallon* burned. More recent designs, however, have improved emissions in diesel cars.

In terms of operating characteristics, diesels have a few quirks that historically have made them relatively scarce in the passenger car market (at least in the U.S.—diesel cars have had more success in Europe). Because they rely on spontaneous combustion, they can be difficult to start in cold conditions. The air coming into the cylinder is not heated, and what little heat is created during combustion is quickly lost to more incoming cold air. For this reason, many diesel designs use electric heaters to either warm the cylinders (with so-called "glow plugs") or the entire engine block in order to ease starting.

As a fuel, diesel also competes somewhat with fuel oil (used in home heating) in terms of how much a given refinery can make from a barrel of crude oil going in. As a result, the price of diesel at the pump tends to rise in winter as more of the crude oil supply is being used to make fuel oil, whereas gasoline does not have a similar rival in the refining process.

Why Is Diesel More Expensive Than Gasoline?

Diesel fuel is cheaper to make than gasoline—it is, literally, less refined—and historically it has been cheaper than gas. In recent years, however, diesel prices have risen to the point where they are often higher than gasoline.

What is driving diesel prices up?[7] First, there is simple economics—in winter, demand for home heating oil competes with demand for diesel fuel, allowing refineries to charge more for diesel relative to gasoline. Second, as we noted in volume I, refining capacity has not kept pace with rising demand for all forms of petroleum products, gasoline and diesel included. Diesel also suffers from a bit of an inferiority complex in the U.S. as it is regarded primarily as a commercial fuel and is not used for passenger vehicles in nearly the same numbers as in Europe.

Regulation also plays a role. The federal government has put restrictions on sulphur levels in diesel in recent years, which in turn made it more expensive to produce. Then there are taxes. At the time of this writing, federal excise tax on diesel stands at 24.4 cents per gallon, 6 cents higher than the tax on gasoline. It's not hard to see how a "commercial fuel" with merely tens of thousands of users would get heavier tax treatment than a "people's fuel" with millions of users (i.e., voters).

Finally, while the demand for all fuels is increasing, there is an even greater demand for diesel outside the U.S. Diesel already accounts for over 30 percent of new vehicle sales in India, and that figure is expected to hit 50 percent in the next few years. China, too, has developed an insatiable appetite for diesel. So, the supply and demand balance for diesel has combined with a less favorable tax treatment and tightening regulations to push up prices, perhaps permanently.

Natural Gas

Natural gas is a remarkably versatile fuel, and is used in home heating, cooking, and a variety of industrial applications in addition to transportation. It has been found to be one of the most environmentally friendly fossil fuels thanks to its capacity for burning cleanly, and its popularity is growing. Auto manufacturers have produced natural gas driven cars for years, but historically more common applications are found in small vehicles that operate in a relatively small area, such as forklifts and those odd-looking special purpose vehicles scurrying around the tarmac at airports.

All of these vehicles use compressed natural gas or "CNG," which is stored in high-pressure cylinders. Several passenger vehicles are available today that operate on compressed natural gas. Some run on natural gas only and others can run on either natural gas or gasoline (called *bi-fuel* or *flex-fuel vehicles*). CNG is sometimes confused with liquefied natural gas (LNG). Chemically speaking, these two are the same thing, but CNG has a lower cost of production since it doesn't have to be compressed and super-cooled into liquid form. On the other hand, LNG requires a much smaller area to store the same amount of energy—that's why you won't see CNG tankers on the high seas today.

The main drawbacks to CNG as a vehicle fuel are range and safety. The energy density of CNG is a fraction of that for gasoline, and the fact that the fuel is stored under high pressure (around 3,000 psi) creates an inherent hazard for a motor vehicle. Still, for shorter-range vehicles like those mentioned earlier, it is a viable solution being used all over the world today.

How Does CNG Compare to Gasoline as a Fuel?

CNG has a number of big pluses as a motor fuel. It can be used in gasoline vehicles and even modified diesel engines, it's clean burning, and it can be found in a wider variety of geographies than oil. The main drawback with CNG is operating range. Large storage tanks can be integrated into automobile designs to minimize their impact on storage space, but there's no getting around the fact that CNG vehicles currently only go about half as far as comparable gasoline powered cars.

Then there is the question of refueling. The amount of time it takes to fill a CNG storage tank depends entirely on how much pressure the gas is under in the main tank. Under high pressure (say, 3,000 to 4,000 psi), you could expect to fill up in a matter of minutes. At lower pressure, as with compressors intended for home use, it could take hours.

Liquefied Natural Gas

Liquefied natural gas is made by refrigerating natural gas to -260 degrees Fahrenheit to condense it into a liquid. This is called *liquefaction*, and it also takes out most of the water vapor and other impurities at the same time. LNG is usually more than 98 percent pure methane. The liquid form is far denser—both in terms of mass and energy—than CNG. That means LNG overcomes the range limitations of CNG-powered vehicles, though the fuel still has to be stored at high pressure.

Despite these advantages, natural gas powered cars—whether CNG or LNG—still only number around 80,000 in the U.S.[8] Compare that to the 200 million gas and diesel vehicles on the road. Why doesn't natural gas make up a greater percentage of our national fleet? Well, in addition to the range and safety issues noted, there is also the small matter of distribution. Today, there are only 1,100 natural gas fueling stations in the United States compared to over 180,000 for refueling gasoline.

Propane

Propane is usually found in the same geological formations as natural gas, often mixed with it and petroleum. It is chemically similar to both, but propane can do tricks the others can't. You see, under normal atmospheric pressure and temperature, propane is a gas, but if you apply a little more pressure and/or lower temperatures, presto! You've got a liquid. This characteristic has earned propane the nickname "the portable gas."

Thanks to this, propane can be easily stored as a liquid (think barbecue), which takes up much less space than its gaseous form—270 times less to be exact. A thousand gallon tank of propane gas might give you enough fuel to cook with for a week, but the same tank filled with liquid propane would keep you cooking for five years.

Fun Fact: Did You Know?[9, 10]

- There are over 200,000 vehicles on the road today that run on propane.
- There are more than 3,000 propane fueling stations in the United States.
- The cost of a gasoline-gallon equivalent of propane is generally less than that of gasoline.

You can go to a website sponsored by the U.S. Department of Energy to find propane fueling stations across the country. Visit http://www.afdc .energy.gov/afdc/locator/stations/.

Propane is a member of the liquefied petroleum (LP) gas family along with butane, ethane, and pentane. Propane is used so broadly that it's often referred to as "LP gas." So why isn't propane used as a transportation fuel more often? The main reason is simply that it's not as conveniently available as gasoline. It also requires engine modifications, and the cost of converting a gasoline engine to use propane is often prohibitive. Finally, there is a slight drop in miles per gallon fuel efficiency versus gasoline.

Fuel Efficiency Standards (CAFE)

Browse the section titled "Coffee Education" on Starbucks' website to get a sense for how the Seattle coffee retailer has redefined every aspect of the coffee experience. The company has popularized an entirely new vocabulary replete with definitions for minutiae associated with a cup of joe, from how long the coffee beans and water should sit in direct contact with each other during the brewing process to tips on how to measure and assess the aroma, acidity, body, and flavor. For better or worse, the humble daily ritual of grabbing a cup of coffee has become near-synonymous with the retail chain's branded piece of rolled-up cardboard.

What we are about to discuss in this section, however, are the *other* CAFE standards, the ones that pertain to energy efficiency and fuel economy in passenger vehicles. But what on earth is "CAFE" and why should we care?

CAFE stands for *Corporate Average Fuel Economy* and is the sales-weighted average fuel economy of a given manufacturer's fleet of passenger cars and light trucks (i.e., pickups, vans) built for sale in the U.S. in any given year.[11] OK, that's a mouthful, but don't worry, the basic idea here isn't that hard to grasp.

CAFE is simply the average mileage of a given manufacturer's overall line of cars and trucks. It looks at vehicles that weigh in at less than 8,500 pounds, with some fancy math in there to allow for the different combinations of large and small cars from one company to another. The weight limit was intended to exclude large commercial and farm vehicles from the standards, though several large SUV models also fall into this category.

Federal law in 2009 said that all manufacturers must have a minimum CAFE of 27.5 miles per gallon for cars and 23.1 mpg for light trucks. The Energy Independence and Security Act of 2007 changed the weight restriction so that beginning with the 2011 model year, all vehicles under 11,000 pounds will be subject to CAFE regulations. It also sets escalating minimums for trucks, with the figure for 2010 models being 23.5 mpg.

Carmakers must offset sales of vehicles with lower fuel economy by selling more high fuel economy models to get the average for the company's entire production to the magic numbers of 27.5 and 23.5 mpg. If they don't meet these minimums, they have to pay a financial penalty. Conversely, manufacturers who exceed their requirements by improving fuel economy across the board earn credits which they can bank for up to three years to offset any later shortfall in standards. CAFE has special provisions to account for alternative and dual fuel vehicles, including hybrids and flex-fuel vehicles.

Key Concept: CAFE Opens for Business

The Energy Policy Conservation Act was passed shortly after the 1973–1974 Arab oil embargo with the goal of doubling fuel economy in automobiles by 1985 and reducing our country's dependence on foreign oil. The law was enacted by Congress in 1975 and it established CAFE standards for passenger cars and light trucks. The goal was to create a level playing field among all car manufacturers while setting a high (yet achievable) bar for increasing fuel economy across the board. Automakers responded, and within a few years, Americans' average fuel efficiency rose rapidly. Then the bottom fell out of the oil market and fuel efficiency remained essentially flat for the next twenty years.

The CAFE program has been controversial since its inception. Some argue that the standards are outdated and need to be significantly increased. Some others argue that CAFE has failed dramatically.

Advocates of CAFE will tell you that the standards have substantially improved fuel efficiency, and it's hard to deny given the numbers of actual vehicles on the road. Estimates have placed the current national annual gasoline consumption at 14 percent lower than if the standards had not been introduced. On the other hand, these improvements have come at a cost in terms of higher vehicle prices for consumers and lower profits for manufacturers.

Critics will also point to an increase in overall traffic fatalities following the adoption of CAFE. This argument says that the standards have forced carmakers to cut weight from their vehicles, making them less safe. Some lighter vehicles have higher fatalities in collisions with heavier vehicles. Further, any increase in CAFE standards will force manufacturers to put still more light vehicles on the road, thereby increasing the chance of a traffic fatality.

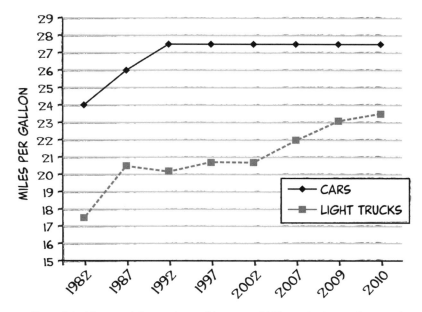

Figure 8.1 After remaining unchanged for years, CAFE standards were increased
under the Energy Independence and Security Act of 2007. (NHTSA)

Critics of CAFE also object to certain aspects of the rules themselves, such as the exemption of heavier vehicles from the standards, which they say undermines the effectiveness of the program.

Allowing carmakers to trade CAFE credits among one another might allow more improvements to be made at a lower cost. Car companies that innovate to improve fuel efficiency could then sell their credits to another manufacturer for a price, and take that savings to the bottom line. Today, CAFE does not allow for buying and selling of credits.

Alternatives to CAFE standards have been proposed that range from increasing fuel taxes to setting up cap-and-trade laws designed to limit carbon dioxide emissions from tailpipes. (We explain the cap-and-trade programs in more detail in Volume 1.)

Who Calculates Fuel Economy and How Is It Done?

The procedures used to gauge fuel economy are set by the Environmental Protection Agency. This agency was created in 1970 as an independent

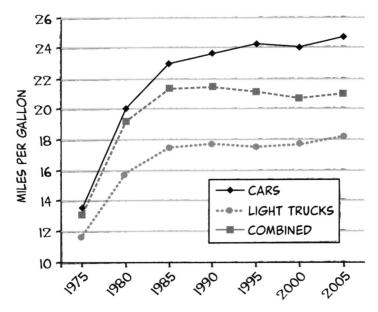

Figure 8.2 Fuel economy across all vehicles has improved only slightly in the past twenty years. (NHTSA)

regulatory body responsible for implementation of federal laws aimed at protecting our environment.

The EPA has long been criticized for using outdated test procedures for measuring fuel economy in cars, and indeed it is a rare case where an individual's actual mileage matches that advertised on the window sticker.[12] Why is this so?

Mainly, it's due to a host of 1970s-era assumptions made by the EPA when CAFE was conceived that don't account for how people actually drive. Use of air-conditioning, driving uphill, cold conditions and traffic don't exist in the world of EPA tests.

In 2006, the EPA proposed adding test situations that matched real-life situations like high-speed driving, rapid acceleration, and stop-and-go traffic.[13] This new system was first applied to 2008 models and was expected to reduce mileage ratings for many models by as much as 20 percent for city driving.[14]

More information on EPA's ruling can be found at http://www.epa .gov/fueleconomy/ and more information on fuel economies of vehicles can be found at www.fueleconomy.gov.

Biodiesel 9

WHILE THERE REMAINS SOME DEBATE about the particulars, virtually no one now contests the fact that fossil fuels are a finite source of energy and that someday we will indeed run out of them. Long before that, the price will likely become so high as to make oil impractical as a transportation fuel. This fact, reinforced recently by rising prices for crude and combined with global warming concerns, has led to increased interest in alternative fuels.

"Alternative" fuels are generally understood to be renewable. The term "biofuels" refers to alternative fuels derived from crops or animals. There is a certain irony here in that oil is perhaps the ultimate biofuel, being derived from ancient plant matter, but it's not really renewable on anything but a geologic timescale. You wouldn't want to wait a billion years for your next fill-up.

Two alternative fuels have emerged as the frontrunners in the biofuel market: ethanol and biodiesel. The first of these is perhaps more familiar to most of us, and we'll cover that in the next chapter. But what, exactly, is biodiesel? The National Biodiesel Board provides the following definition:

> Biodiesel is defined as mono-alkyl esters of long chain fatty acids derived from vegetable oils or animal fats, which conform to ASTM D6751 specifications for use in diesel engines.[1]

Got that? Yeah, we didn't either. The term "biodiesel" in popular usage refers to fuels used in diesel engines but that don't come from petroleum. There is a distinction to be made, though. Strictly speaking, "biodiesel" as defined by the EPA and other regulators must adhere to certain standards, and the used vegetable oil from your local burger joint won't meet them. This so-called simple vegetable oil (SVO) can run diesel vehicles—it just isn't legal to sell as a motor fuel, and it can cause problems in unmodified engines.

You may have seen news stories about "french fry" cars running on discarded vegetable oil or even animal fats from restaurants, but these are typically hobbyists who make their own fuel. While it is relatively easy to convert waste oils like these, there just isn't that much to be had. We eat a lot of french fries, but not anywhere close enough to meet the daily requirements of our diesel-powered cars and trucks.

Commercial biodiesel is made through a chemical process that separates the glycerin from the fat in the oil. This process is called *transesterification* and at the end of it you wind up with methyl esters (i.e., biodiesel) and glycerin, which can be sold for use in a variety of other products, like soap.

Where Does Biodiesel Come From?

If you like Japanese food, you're probably already acquainted with the primary source of biodiesel.

Edamame are soybeans boiled in the pod, and they're just one of a multitude of things made from this remarkable plant. Tofu, soy milk, soy sauce, soy nuts, soy ice cream—the list goes on, and we haven't even left the kitchen. Soybeans are also the source for most of the biodiesel sold in the U.S.

Like petroleum diesel, biodiesel is used in compression-ignition (diesel) engines. Today, it is typical for biodiesel to be blended with regular petroleum diesel to allow it to be used in conventional diesel engines. A blend of 20 percent biodiesel with 80 percent petroleum diesel is called "B20." A 50-50 blend is called "B50," and so on.

B100 biodiesel may require certain engine modifications and is generally not suitable for use in low temperature conditions due to its thick consistency. Biodiesel acts as a lubricant as well as a fuel, increasing the engine's ease of movement as it runs. In fact, the U.S. Department of Transportation estimates that a 1 percent blend of biodiesel could increase "fuel lubricity" as much as 65 percent.

That sounds like a good thing, but biodiesel is also a solvent. It will eventually break down rubber components (e.g., fuel lines). Also, if you start running B100 in a diesel engine that has only ever seen petroleum fuel, pretty soon your fuel will start loosening deposits left behind by the petroleum fuel, potentially causing clogs. Biodiesel suppliers recommend changing the car's fuel pump shortly after making such a transition from conventional fuel to biodiesel. On the other hand, if you run a diesel engine on B100 from the beginning, the fuel will leave no deposits behind, extending engine life.

Using biodiesel also cuts down on emissions across the board (unburned fuel, carbon monoxide, sulfates, particulate matter), and the more you use the greater they come down. CO_2 emissions are also reduced using biodiesel—by 75 percent using B100 and 15 percent using B20.

Biodiesel also offers a few other notable benefits. For one, it's biodegradable and breaks down four times faster than petroleum-based diesel. It also has a much higher flashpoint—300° F compared to conventional diesel's 150°F—so it's less likely to explode when exposed to high temperatures.

Where Can I Buy Biodiesel?

Biodiesel is available in all 50 states in the U.S., and the National Biodiesel Board maintains an up-to-date list of retail locations at www.biodiesel .org. The website also has a handy online tool to help you determine the reduction in your emissions footprint by switching to a biodiesel blended fuel from petroleum diesel.

If you live in the South, you might see signs for a particular brand of biodiesel called "BioWillie." It's made by a company bankrolled by country music legend Willie Nelson that tries to support family farms with a new market for their soybean crop.

Large Scale Economics of Biodiesel: Can It Replace Fossil Fuels Entirely?

Biofuels, and biodiesel in particular, are very attractive because they can be made from renewable, natural sources right here in the U.S. The $64 question, then, is this: can biodiesel and biofuels generally replace gasoline and diesel in motor vehicles?

The answer is along the lines of, "yeah, well, sort of, but not yet." If you took all of the arable land in America—that's around 470 million acres—and put it all into soy production, you'd be able to produce just about enough biodiesel to replace conventional diesel on the road.[2, 3] So, while biodiesel offers many advantages, our current methods of production are simply not up to the task of knocking oil off of the transportation throne.

Recently, however, research has begun on using algae to produce biofuels. This prospect is exciting for a few reasons. First, algae can be grown pretty much anywhere, eliminating the issue of taking farmland out of food production to grow plants for fuel. The potential yields are also much higher than with soybeans. One estimate indicates 15,000 square miles (i.e., the state of Maryland plus a couple thousand square miles), would be enough in theory to produce enough algae-based biofuels to fully displace diesel.[4]

Ethanol   **10**

Introduction

We've all heard the phrase "alcohol and driving don't mix." People who get behind the wheel after consuming alcohol can get into a lot of trouble. Even if they manage to avoid harming other people, vehicles get impounded, licenses get suspended, and even the occasional hotel heiress goes to jail.

But what if the car is the one consuming the alcohol?

Ethanol, the very same alcohol found in everything from vodka martinis to mai tais, is an increasingly popular fuel for automobiles. Ethanol has been in the news lately but in fact Henry Ford touted ethanol (or ethyl alcohol) as the "fuel of the future" way back in 1925.

"The fuel of the future is going to come from fruit like that sumach out by the road, or from apples, weeds, sawdust—almost anything," he said.[1] "There is fuel in every bit of vegetable matter that can be fermented. There's enough alcohol in one year's yield of an acre of potatoes to drive the machinery necessary to cultivate the fields for a hundred years."

When Ford designed the Model T, he fully expected ethanol to become a major automobile fuel. It was easy to make, safe to handle and aside from some difficulty starting on cold mornings, it worked pretty well. However, as we saw in chapter 8, gasoline emerged as the dominant transportation fuel in the early twentieth century in large part due to the growing supply of inexpensive petroleum from domestic and foreign sources. Then came the oil crisis of the early 1970s and ethanol got a new lease on life.

The Ethanol We're Already Using

In the U.S., we now consume more than 15 billion gallons of ethanol-blended gasoline a year, totaling 12 percent of all fuel sales.[2] In the United States each year, approximately 2 billion gallons of ethanol are added to gasoline to increase octane and improve emissions.[3] There's a good chance you are using ethanol already without even knowing it. Many states in the U.S. blend small amounts of ethanol with gasoline to reduce the emissions the vehicles produce. The typical blend of ethanol and gasoline is referred to as E10 (as in 10 percent ethanol, 90 percent gasoline) but other blends with a higher percentage of ethanol such as E85 and even E95 are being tested. Those little bottles of "fuel system cleaner" people add to their tank are also mostly alcohol (isopropyl alcohol to be exact).

Fun Fact: Which States in the U.S. Have Mandated Blending Ethanol into Gasoline?

Nine states have Renewable Fuels Standards that require the use of ethanol-blended fuel:[4] Hawaii, Iowa, Kansas, Louisiana, Minnesota, Missouri, Montana, Oregon, and Washington. California does not require renewable fuels, but the state does have an oxygen standard, and the only substance for adding oxygen to automotive fuels that is approved by the California Air Resources Board is ethanol.

Fourteen states don't require ethanol blending but have some type of retail incentives for it, whether for E10, E85, or both types of ethanol-blended fuel. They are Alaska (E10), Idaho (both E10 and E85), Illinois (both), Iowa (both), Kansas (E85), Maine (both), Missouri (both), Minnesota (E85), Oklahoma (both), South Dakota (both), Connecticut (both), Hawaii (both), South Carolina (E85) and Alabama (E10).

Making Ethanol

Ethanol can be made from any plant source that contains sufficient amounts of sugar (e.g., beets, sugar cane) or materials that can be converted into sugar such as starch or cellulose (e.g., corn). Cellulose, which is found in many grasses and woody plants, can be converted to sugar, but the process is more complicated and expensive than starting out with sugar.

The typical ethanol production process starts by grinding up a feedstock, which in the U.S. is usually corn. The starch is then converted into sugar, which is used to feed microbes that metabolize it into ethanol and carbon dioxide.

Ethanol contains 35 percent oxygen. The more oxygen present in combustion, the more completely the given fuel will burn, and that means less emissions. Not surprisingly, then, ethanol is often used as an oxygenate in gasoline blends as an alternative to toxic additives such as benzene, a known carcinogen.

Ethanol is nontoxic, water soluble and it breaks down quickly. It poses no threat to surface water and ground water. When a gasoline spill contaminates soil or water, the ethanol is the first component to break down.

Ethanol also reduces particulate matter formation by diluting aromatic content (e.g., benzene or other octane enhancers) in gasoline. These substances may help your car go faster and allow you to tap into your inner Cole Trickle (Tom Cruise's character in *Days of Thunder*, remember?), but unfortunately aromatic additives have serious health effects ranging from damaging your bone marrow to depressing your immune system. That's one kind of "aromatherapy" you don't want any part of.

Biofuels' Potential

According to a study by the National Resources Defense Council (NRDC), increased use of biofuels could deliver a range of benefits: cutting our transportation-related greenhouse gas emissions by 80 percent by 2050; providing a new source of revenue for farmers, to the tune of $30 billion a year by 2050; and putting unproductive farmland back into service with a minimum of requirements for water, fertilizer, and so on. The NRDC further asserts that if production ramps up, biofuels can be cost-competitive with petroleum-based fuels.[5, 6, 7]

That's a big "if," to be sure. It has been done to some extent in other countries and we'll talk about one in a moment, but it's probably best to view the rise of biofuels as supplemental to gasoline and diesel, at least for now. Displacing those two outright will take decades, and will probably involve a range of alternatives (e.g., electric vehicles) in addition to ethanol and biodiesel.

Flex-Fuel Cars

Think of the flex-fuel cars program as a low-fat diet for vehicles. Running cars on carbohydrates (ethanol) instead of fossil fuels is the automotive version of consuming "good carbs" and staying away from "bad fat."

Flexible-fuel vehicles can run on pure gasoline, but are also capable of using up to 85 percent ethanol (E85). The only difference is that the more ethanol there is in the mix, the lower the miles per gallon. Running E85 will mean a 20 to 30 percent reduction in mileage. This is due to the difference in energy density between gasoline and ethanol, which we discussed in chapter 8.

Flex-fuel vehicles look the same as regular cars, and many models are sold in the U.S. Perhaps you even own one already.

Ethanol and Global Warming

Compared to gasoline, ethanol is a much cleaner burning fuel. It releases less carbon dioxide into the atmosphere than fossil fuels. When blended with regular gasoline as an octane enhancer, it can cut emissions by over 50 percent. The added oxygen it offers also reduces carbon monoxide emissions by 25 to 30 percent, according to the EPA, and it cuts down on emissions of sulfur dioxide (think acid rain).

All of these virtues aside, however, there is a debate about ethanol's environmental credentials in terms of something called *net energy*.[8] In other words, does it take more energy to produce ethanol than it delivers as a fuel? This issue is especially controversial in the United States as most ethanol here is made from corn, which using conventional farming practices requires a good deal of energy to grow. Plus, corn only gets you to starch that must then be converted to sugar before you get to ethanol, and that in turn requires more energy. So where does that leave us?

On the "pro" side of the argument, there are a number of points in favor of ethanol. First, simply by replacing gasoline with ethanol one can reduce carbon dioxide emissions by up to 29 percent. There is also the concept of the carbon cycle. Being derived from plants, ethanol essentially soaks up CO_2 during its "production" as growing corn stalks before releasing it as a motor fuel. The next generation of corn then grows using the CO_2 from the last batch's life as ethanol. Some studies show that this cycle is cumulatively positive over time, meaning more and more CO_2 is being consumed than is put back into the atmosphere.

The U.S. Department of Agriculture performed a study that showed ethanol delivers two thirds more energy as fuel than it required to be produced.[9] Those impressive results were aided by a variety of factors such as no-till farming practices, higher yielding varieties of corn and generally more efficient processing techniques. The USDA further estimates that each BTU (British thermal unit) used to produce a BTU of gasoline could make 8 BTUs of ethanol.

So all of this sounds pretty good, right? The problem, ethanol critics will tell you, is that modeling ethanol production and predicting net

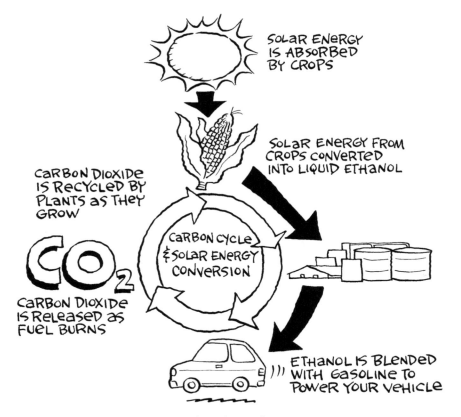

Figure 10.1   The carbon cycle. (Doug Jones)

energy are highly complex undertakings, and that there is a great deal of variation in actual practice.[10] For example, the "best farming practices" are not used in the majority of cases today. Corn-based ethanol in particular comes under scrutiny because of its lower net energy compared to ethanol produced from other feedstock like sugar cane that can skip the starch-to-sugar conversion step.

Also, agricultural systems release methane from the soil, which is more that 20 times as powerful as carbon dioxide as a greenhouse gas, and nitrous oxide, which is 300 times more powerful. Still, the benefit of no-till practices or the use of perennial crops to mitigate the release of these gases is short-lived. After about 20 years, the soil reaches a new equilibrium and will no longer absorb as much.

So, like so many other questions we've endeavored to answer, the bottom line on the net energy of ethanol comes down to "it depends."

Brazil's Sweet Success: An Ethanol Story

What comes to mind when you think of Brazil? Carnival? Mojitos? Whatever might have popped into your mind, it's probably not "energy independence." OK, maybe it was, but moving on . . .

Motor fuel sold in Brazil is at least 24 percent ethanol, and much of it is 100 percent ethanol. Today, more than 70 percent of new cars produced in Brazil are "flexible-fuel" vehicles that can run on either ethanol or gasoline.[11] So how did this one country achieve such remarkable progress in moving away from fossil fuels?

Brazil's ethanol program originated in the 1970s in response to the Arab oil embargo and global uncertainties of the oil market. Yet countries wanting to follow Brazil's example may be leery of following their method to a T because the road to the current state of affairs in Brazil is as bumpy as a jungle path.

The first global oil shock in 1973 hit Brazil hard. At the time, the country imported over 80 percent of its fuel, and within months, the quadrupling of oil prices pushed the Brazilian economy into recession.

Within two years, military and civilian leaders responded by laying the groundwork for a robust alternative fuel industry, mandating ethanol use, dictating production levels and establishing prices. The government started lending money at low interest rates for the construction of ethanol processing facilities. Brazil also funded research to develop a car that would run exclusively on ethanol.

Just as these investments were beginning to pay off, in 1986 the price of oil plunged. Brazil's economy then experienced a period of spiraling inflation that effectively killed off ethanol subsidies. To make matter worse, sugar cane producers started getting a lot more business from sugar mills due to a runup in the global price of granulated sweetness. The result was less sugar going into the ethanol industry, which created fuel shortages. You can imagine how public sentiment toward ethanol was affected.

By the early 2000s, ethanol-only cars in Brazil dropped to around 3 percent of the market and it looked like the great experiment would fail. But decades of groundwork and investment in the infrastructure kept the ethanol industry alive. Producers refined their processes and lowered costs. Gas stations still offered ethanol, which continued to be taxed at a much lower rate than gasoline.

As gasoline prices began to rise again—this time to stratospheric heights—ethanol rebounded and is available today at 29,000 gas stations nationwide. Compare that to the roughly 600 of America's 180,000 stations. Ethanol accounts for about 20 percent of Brazil's transportation fuel market, compared to a global average of 1 to 2 percent.

Currently, ethanol production reduces petroleum imports to Brazil by about 200,000 barrels a day. Some estimates have put the avoided cost of petroleum fuels since the beginning of Brazil's ethanol experiment at $120 billion. Ethanol has grown beyond the ground transportation industry in Brazil, and is now being used in the aviation industry in small planes such as crop dusters.

So how does the evolution of the ethanol industry in the U.S. compare to that of Brazil? Can't we just take a page from their playbook and start making cheap ethanol here?

Not so fast. Brazil has a few important things going for it. First, as we noted earlier, it's easier (and cheaper) to make ethanol from sugar than it is from starch, and Brazil has a lot of sugar. The national liquor is cachaca (pronounced ca-SHA-sa), which is fermented cane juice. Growing sugar cane in Brazil is less expensive too, thanks to plenty of land, adequate rain, and cheap manual labor that all help to keep costs down. By contrast, corn-based ethanol comes with a 30 percent premium over sugar-based fuel, and that only accounts for the differences in processing—labor and other costs aren't even factored in.

Still, Brazil has demonstrated that ethanol is a viable alternative to gasoline, and the rest of the world has taken notice. India is second only to Brazil in sugar production, and the governments in nine Indian states passed ethanol mandates in 2003 that call for gasoline to be blended with 5 percent ethanol ("E5" you might call it). India is also exporting ethanol to places that can't produce it as economically. Japan, for example, is considering a 3 percent blending rule that on its own could translate into a 30 percent increase in global ethanol demand.

Presently the global ethanol market is embryonic compared to that for the established fossil fuels. But with the converging pressures of climate

---

Fun Fact: U.S. Trade Barriers to Importing Ethanol

The U.S. has set up a 54-cent-a-gallon tariff on imported ethanol designed to protect domestic producers who rely overwhelmingly on corn as a feedstock.

change and energy security, many nations are already looking at ethanol as part of the solution.

Backyard Ethanol

Want to make your own ethanol? We have good news for you. You don't have to be a big corporation or the next John D. Rockefeller to make your own fuel. You can make it yourself in your very own backyard! Type in "make your own ethanol" on your favorite search engine and it will return links to a handful of entrepreneurs who sell stills or still-building kits to make ethanol in your own backyard.

A typical process is one where yeast, sugar, corn, and water are mashed together and left to ferment for two days or more. The alcohol that is produced during fermentation is isolated using simple distillation. Just bring the mash to a boil and capture the vapors—when they cool, they'll condense into a liquid, ethanol.

If you do embark on this adventure in home chemistry, don't forget to denature the end product by adding the appropriate ingredients to make it inconsumable by humans. Of course, if you skip the denaturing you would find yourself with something else, 180- to 190-proof moonshine. Remember, it's against the law to manufacture your own alcohol without a permit, so be sure to get a "small fuel producer" permit from the U.S. Alcohol and Tobacco Tax and Trade Bureau (TTB) or you could find yourself in the slammer before you can say Cheers! Salud! or Kampai!

Check out the following website before you start your own ethanol brewery: http://www.ttb.gov/tax_audit/permits.shtml.

Cellulosic Ethanol: Panacea or the Real Thing?

Once upon a time there lived two brothers named Grimm, and one of their fairy tales involves a poor miller who had a beautiful daughter. In a boastful moment, the miller tells the king that his daughter has a unique ability to spin straw into gold. The king calls for the girl, shuts her in a tower room with straw and a spinning wheel, and demands that she start spinning out the 24K twine or be locked up in the dungeon to die.

The story of Rumpelstiltskin isn't so far removed from the world of ethanol. Surely the fuel is valuable, but it's not exactly easy to produce and comes with some baggage (see the food vs. fuel debate in this chapter). But there is another way to make ethanol that avoids these nagging issues and can even play a constructive role in land use. Cellulosic ethanol is "the other ethanol," produced from all kinds of plant matter, even organic refuse. And boy, do we make a lot of refuse. Fifty years ago the average American generated 2.7 pounds of trash a day. Today, the average is 4.5 pounds.

Some of this waste undoubtedly can be recycled, and some is. Some of it can be burned in a waste-to-energy power plant to generate electricity, and some of it—the plant matter—can be used to make ethanol.

Cellulosic ethanol is produced from the stalks and stems of plants as opposed to sugars and starches as with sugar cane, beets, corn, and other typical feedstock. Making ethanol from grass, agricultural waste, and forestry waste, in theory, would not be as expensive and energy-intensive as farming corn. Plus, economists say that forests (especially the plant waste) don't compete for land with food crops like corn, rice, and wheat. Still, cellulosic ethanol presently costs around $2.20 a gallon to produce, roughly double the price of corn ethanol.

There are a few approaches to manufacturing cellulosic ethanol. Some apply heat and pressure to the waste wood and grass to turn them into a gas, and then apply a chemical catalyst to convert the gas into alcohol vapor, which is then condensed into the liquid fuel. Others employ a process that uses enzymes to convert the agricultural waste directly to ethanol. To the extent that these new processes require less fossil fuel to create the ethanol, they can even have a bigger effect in reducing greenhouse-gas emissions than corn ethanol.

Washington's Role

In his 2007 State of the Union address, then-President Bush unveiled his plan for strengthening America's energy security. Dubbed "Twenty in Ten," the plan called for reducing U.S. gasoline usage by 20 percent in the next ten years. Among other things, it set a mandatory production target of 35 billion gallons of alternative fuels in 2017. If achieved, this would displace 15 percent of expected annual gasoline use.

As a point of comparison, the Energy Policy Act of 2005 set a target of 7.5 billion gallons of renewable fuels by 2012, a benchmark that we've already surpassed. The current ethanol production capacity in the U.S. is more than 11 billion gallons a year. But according to industry experts and environmental groups, getting from there to 35 billion gallons will require loan guarantees and other government incentives.

American farmers have explored new markets for alcohol-based fuel four times in the past 125 years. The first was around 1906; the second time was in the 1930s—with Henry Ford's blessing and support. The third was during the oil embargo of the 1970s and the fourth was in 2007 with crude oil prices on their way north of $100 a barrel. Since then, oil has plummeted to $40 a barrel and returned to its resting point as of this writing at around $85 a barrel. Will this boom and bust cycle in oil crash the global ethanol industry like it did in Brazil? That probably depends more than anything else on government policy and incentives until another rise in oil prices reignites the search for alternative fuels.

What Is the Food vs. Fuel Debate?

Quick, think of a food that says "summertime." We're betting that some of you thought of corn on the cob, and with good reason. Corn is steeped in America's history—culinary and otherwise—but it has now become the very bedrock of our food industry. Call it a "superstaple." If you were to look at the ingredients of every single item in your grocery store, you would find corn or some derivative of it on the list of one out of every four packages.

In addition to the snack food aisle, corn has found its way into beer, vitamins, soaps . . . the list goes on. Even the aspirin you pop to relieve a headache has corn in it—starch used to make the film that covers the tablet.

So what's the problem?

Today, ethanol in the U.S. is made mostly from corn, and the price of ethanol as a fuel can be very unpredictable and volatile (as all fuel prices are volatile and unpredictable). Simple economics states that the use of corn for fuel *in addition* to all those other uses makes corn a more precious commodity. The more precious a commodity, the pricier it gets (remember the craze for "retired" Beanie Babies back in the 1990s?). If the miracle grain that winds up in 25 percent of our food products goes up in price, there's a good chance your grocery bill will too, depending on what you buy.

Corn is also widely used as feed for livestock, so an increase in corn prices will impact the cattle rancher and hog farmer too. The result, according to some, is that a substantial increase in ethanol production will drive up prices in food due to the increase in the price of corn, not to mention the potential for corn to displace other food crops in the field and thus drive up their prices as well.

For their part, ethanol producers dismiss such claims as propaganda. As an example, a six-pack of soda contains about 6 cents worth of corn sweeter. A $5 bucket of movie theater popcorn contains just 0.15 pounds of corn before popping. Based on an average 9-cents-per-pound rate that farmers made on the corn, that bucket contains slightly more than a penny of popcorn. Higher corn prices means the moviegoer may now have to pay about 2 cents of popcorn per bucket, instead of one.

While corn prices certainly have an impact on food prices, we'd humbly suggest another fuel be considered in this debate: oil. Our entire food production and delivery system is dependent on oil and while we haven't run the numbers, it wouldn't surprise us at all to learn that an increase in the price of oil will have a far greater impact on food costs than an equivalent increase in the price of corn.

Methanol

While ethanol gets all the attention, we can't overlook another member of the alcohol family in our discussion of alternative fuels. Methanol (aka "wood alcohol") can be made from plant material, and it can even be derived from coal. Presently, the U.S. produces one quarter of the global supply of methanol, mostly by deriving it from natural gas, though it is hardly used in vehicles outside of racing circles.

Methanol is separated chemically from ethanol only by a few atoms (two hydrogen, one carbon, in case you're interested), but unlike its socially effervescent cousin, methanol is very poisonous. As a motor fuel, it also has a few drawbacks. One is that it is corrosive to some metals, including aluminum. As a result, conventional gasoline engines must be modified to burn it. After all, it's not so good for your fuel to eat your car.

Also, due to its toxicity and solubility in water, there is a considerable risk of ground water pollution if methanol were to be adopted on a wide scale. Just look at what's happened with MTBE, once a "good guy" additive used to reduce emissions and now a "bad guy" thanks to leaking underground tanks that contaminated well water.

Hybrids, Plug-Ins, and Electric Vehicles **11**

B Y NOW, WE'RE GUESSING THAT you've probably heard of hybrid cars. Indeed, these vehicles have arrived—in more ways than one. For example, what do the Toyota Prius, a cup of Starbucks coffee, the Motorola RAZR phone, and McDonald's fries have in common? They are all tokens in the Monopoly board game's "Here and Now" edition, which replaces the traditional game pieces with more recognizable twenty-first-century icons.

During the highly publicized California recall election for governor in 2003, an unexpected rivalry emerged between actor-turned-candidate Arnold Schwarzenegger and political commentator-turned candidate Arianna Huffington. After learning of the now-governor's declared candidacy, Ms. Huffington observed, "This race will be the Hummer versus the Hybrid."

Cynics everywhere were rubbing their hands with glee when Al Gore's son was caught in early 2007 speeding at over 100 mph driving a Toyota Prius with a cocktail of prescription drugs on board, as they hoped to tarnish Gore's whiter-than-white (or should that be greener than green?) reputation. But ironically, the bigger response to the story has been "A hybrid can go *that* fast? Really?"

In this chapter we will give you the scoop on hybrid cars, but first we need to spend a little time on electricity and that's why we've lumped all-electric vehicles in here with their hybrid and plug-in cousins.

Electricity in Automobiles

Electricity has been used to power cars for well over a century. Electric vehicles (EVs) work by storing energy from an external source (e.g., the power grid) as chemical energy inside of a battery. The energy is released—or discharged—from the battery to run an electric motor.

In the late nineteenth and early twentieth centuries, it was the battery-powered electric vehicle that came to the rescue of ordinary citizens who were at that time reliant on horses and horse-drawn buggies for short-distance personal transport. In fact, battery electric vehicles even predate gasoline-powered vehicles.

Ironically, electric motors contributed to the fall of the very cars they powered.[1] In 1913 Cadillac introduced a gasoline-powered car with an electric starter, thus relieving drivers of the onerous and sometimes even dangerous task of crank-starting their engine. That gave the internal combustion engine vehicle an advantage, and with the abundance of cheap gasoline to fuel it, the car design we know so well today pulled away from its electric forerunner.

Electric vehicles do not produce any tailpipe emissions since they don't have a tailpipe. They can be up to 99 percent cleaner than conventional vehicles, even when you take into account the emissions of the power plants that generate the electricity used by the car. Even if your EV is recharged using coal-fired power, it can still cut global warming emissions by as much as 70 percent over gasoline-powered vehicles.

The first-generation EVs had a real-world driving range of 50 to 80 miles. Advances in battery technology now permit electric cars to travel well over 200 miles on a charge. Electric vehicles also have a few interesting characteristics. They are very quiet, and at rest make no sound at all. Electric motors require no gears and deliver all of their available torque immediately, even from a standing start. That allows them to accelerate far more rapidly than conventional cars. In racing jargon, they are "fast off the line," and if you want proof just have a look on YouTube for videos showing a drag race between a General Motors EV1, a Mazda Miata, and a Nissan 300Z. We'll be nice and just say the gas burners didn't win and leave it at that.

So, why aren't electric vehicles used more broadly today?

EVs (also known as battery electric vehicles or BEVs) made a comeback in the 1990s following the California Air Resources Board's (CARB) requirement for a minimum number of zero-emission vehicles to be introduced in the state every year.[2] However, the EVs died a second death after CARB withdrew their requirement following complaints by auto manufacturers that the requirement was economically unsustainable due to insufficient demand. The rise and fall of General Motors' EV1 is the subject of the 2006 film, *Who Killed the Electric Car*. In an ironic twist of events, more than a decade after the death of the EV1, General Motors announced the 2010 launch of the all-electric Chevrolet Volt in an effort to reinvigorate its lineup.

The future of all-electric vehicles is still somewhat cloudy at this time. With one or two notable exceptions (see next section), all of the major auto manufacturers put an end to their EV production lines. However, there is no denying that the development of the electric vehicle has made important contributions to advancing the drive train and energy storage technologies that are behind today's hybrid vehicles.

Will all-electrics make another comeback? It may already be underway. Nissan recently introduced the Leaf, an all-electric vehicle that with U.S. tax incentives will sell for between $20,000 and $25,000. Other carmakers are working on all-electric models, too. Still, there are issues of cost (the batteries aren't cheap) and driving range that could put a drag on the EV's glorious return.

Tesla Motors

Let's say you're an Internet billionaire. You've built up one of the biggest web businesses on the planet and now you're looking for something to do next. If you're Elon Musk, co-founder of PayPal, you might start a new car company.

Tesla Motors, which takes its name from the nineteenth-century scientist who first advocated use of alternating current (among other things), is producing the world's first modern line of all-electric vehicles.[3] The first model is a low-slung sports car that has the distinction of being faster than just about anything else on the road (zero to 60 mph in less than four seconds). At over $90,000 they're not intended for the masses, but the car has generated a tremendous amount of buzz (some might call it hype). If nothing else, Tesla has reignited the idea of the electric car in the popular imagination as something more than a technological curiosity.

Tesla's next model is a high-end sedan, which as we write is still in development. And a more affordable four-door is up next. The line forms to the left.

What Is a "Hybrid" Anyway?

A hybrid vehicle is simply one that combines two or more sources of power. Most of us think of "hybrids" as cars that run off a rechargeable battery and gasoline as their two sources of power since these have been the type marketed by automakers over the last several years.

Hybrid cars work by splitting the drive duty between an internal combustion engine and an electric motor. The electric motor provides torque at

To Jay Leno, Electric Cars Are "The 100-Year-Old New Idea"

He is very excited about the current interest in alternative fuels for vehicles, but for Jay Leno alternative-fuel cars are yesterday's news. Why? The late-night TV host and avid car collector owns several vintage electric vehicles. In the May 2007 issue of *Popular Mechanics* magazine, Jay writes about his prized EVs. One of them, an Owen Magnetics, with a gas engine and an electric generator, can even be considered an ancestor of today's hybrids—it uses the same principle as the forthcoming Chevy Volt.

But Leno believes that battery technology has not progressed enough in the last 100 years. He is able to get 100 driving miles out of his hundred-year-old Baker Electric vehicle. Not much different from what you can get out of a modern electric car on a charge.

For links to the original article in *Popular Mechanics*, go to http://www .popularmechanics.com/automotive/jay_leno_garage.

low speeds and an extra boost for, say, driving up a big hill. Shifting between the two power sources is automatic; the driver may not even be aware of it except perhaps for the sound of the gas engine kicking in at low speeds.

Most hybrids tend to have better gas mileage in the city since the stop-and-go environment is where the electric motor really shines. They still get good mileage compared to conventional cars at highway speeds, but there the advantages of the electric motor have less of an impact.

This partnership between gasoline and electricity represents the design behind all of the major hybrid vehicles being sold today. The Toyota Prius, the Honda Civic hybrid, the Ford Escape hybrid—all of these use the gas-electric arrangement.

The Chevy Volt operates on the same basic principle—it uses an electric motor and a gasoline engine—except that the electric motor is the primary power source and the gasoline engine is there simply to recharge the battery. This design puts gasoline in a supporting role, drastically reducing the amount consumed per mile. The Volt reportedly will get in excess of 150 miles per gallon.

Plug-in Hybrid Electric Vehicles (PHEVs)

Plug-in Hybrid Electric Vehicles (PHEVs) work just like other hybrids, but they also have the added feature of an extension cord. You can plug them into any power outlet and recharge the battery. The result is that the battery can be used more heavily, thus further reducing the amount of fuel used by the gasoline engine.

Presently, Toyota has plans for the next iteration of the Prius to come with available plug-in capability, but up until now PHEVs had to be made using upgrade kits that cost between $5,000 and $12,000 and voided the manufacturer's warranty.

Batteries

Batteries, as mentioned earlier, store energy in chemical form that can be released on demand as electricity. This electrical power is used by the car's ignition system for cranking the engine as well as powering the lights and other accessories.

Car batteries are usually the lead-acid variety, and they weigh a lot— remember the last time you changed one? But these batteries do not have the strength to move the car. For that, we need a completely different kind of battery technology.

Batteries that are used for the main power source of a hybrid or electric vehicle are called *traction batteries*. The ones in hybrid vehicles are usually nickel metal hydride batteries (abbreviated NiMH), another type of re-chargeable battery. A technology that shows even greater promise is the lithium-ion battery (a.k.a. Li-ion battery). What is so special about this technology?

Li-ion batteries pack a punch. Their energy density is much higher than other battery technologies, which is why they are used in laptop computers and even power tools like cordless drills. These batteries can be formed into a wide variety of shapes and sizes to fit within the available space inside the hood or trunk of a car. Since lithium is the third lightest element, these batteries offer substantial savings in weight compared to other batteries (lead vs. lithium? No contest!).

So where's the catch?

The first issue is durability. Getting to, say, a ten-year lifespan is a high bar for Li-ion batteries to reach (if you own a laptop computer, think of how the battery life degrades over even just a year or two). There's also a question of of safety. Li-ion batteries got some pretty bad press a few years ago when a few of them made smoking embers out of the laptop computers they were supposed to run. Now think of a battery hundreds of times that size—in a moving vehicle—and you can see there might be some concern. There is also the question of what to do with a Li-ion battery once its life in the car is over. Recycling is a possibility, but that also plays into our final issue which is, of course, price. Once the safety, durability, disposal and performance issues of the battery are solved, will it be economical to offer to the market? Getting the technology to work at an affordable price is the rest of the challenge. The major carmakers are placing their bets now, so the next few years should be very interesting.

The Persian Gulf of Li-ion Batteries: Bolivia

To make lithium-ion batteries, you need lithium. Where can you find lithium? Well, it turns out that an astonishing 50 percent of the global supply of the stuff is found in Bolivia, which happens to be the poorest country in all of South America. The metal is actually found in the brine underneath the world's largest salt flat known as the Salar de Uyuni high up in the Andes. It's almost impossible to reach, and there is nothing close to the infrastructure needed to support industrial production.

Still, the Bolivian government is well aware of the potential for this resource to reshape the country. They are engaged in research now to fully understand the processes needed to exploit the Salar de Uyuni before they begin to negotiate with foreign companies that have already begun courting government ministers.

The idea of not just one region but a single country having a lock on what could be a vital resource is daunting to say the least. Battery research and development continues, but Li-ion is for now the most viable technology to support wide scale production of electric and plug-in hybrid vehicles.

Are Plug-ins and Electric Vehicles Really That Clean?

OK, so you're thinking at this point, hybrids and PHEVs are great, even with the technological challenges they face. What's not to like?

The debate typically goes like this: Why spend all those extra dollars to buy an electric car or to convert a hybrid car to a plug-in hybrid when they consume power from polluting coal-fired plants? Won't we have to build more power plants to help recharge all these battery-powered cars? How is polluting the atmosphere with coal any different from polluting the atmosphere with tailpipe emissions?

Thankfully, the verdict is in. Many studies have been conducted by experts in the industry and it's not even close: emissions from hybrids and EVs are far less than their gasoline powered counterparts, even when the dirtiest power plants are used to generate the electricity.

The Pacific Northwest National Lab (one of the U.S. Department of Energy's ten national laboratories) conducted a study in 2007 to understand how many new power plants will have to be built to support a growing base of PHEVs and other electric vehicles. Their findings say that no extra power plants will have to be built as long as cars are charged at night.[4]

The Electric Power Research Institute (EPRI) and the Natural Resources Defense Council (NRDC) recently completed another study to assess the benefits of plug-in electric hybrid vehicles and their impact on greenhouse gas emissions. Among the study's key findings was that widespread adoption of PHEVs could reduce vehicle GHG emissions. If implemented on a wide scale, they can cut those emissions by more than 450 million metric tons annually in the next forty years, and that's equivalent to removing 82.5 million conventional passenger cars from the road.[5]

How is this possible? In short, power plants are much more efficient than automobile engines as far as turning the latent energy in a feedstock (coal, gasoline, etc.) into usable energy. Our electric grid is also getting cleaner over the years as we discuss in volume I. Government mandates regarding the use of renewable energy, limits on emissions and advancing generation technologies are all making the power grid cleaner and more efficient.

Hybrid Technologies Go Commercial

Hybrid cars have become the poster children for energy savings in the automobile world. They are also beginning to appear in other modes of transportation, too.

Commercial vehicles that have long idling times or do a lot of starting and stopping tend to gain the most from hybrid technology. School buses, delivery trucks, and other short-range vehicles are obvious candidates for hybridization. FedEx, for example, is building a fleet of nearly one hundred hybrid delivery trucks. There are even hybrid designs for heavy construction vehicles. A hybrid excavator, for example, can charge its batteries every time the excavator swings to the side using the same regenerative braking technology used in trains and hybrid cars.

Can Hybrids Cut Our Oil Dependence?

In the words of a recent vice presidential candidate, "You betcha!" Think of hybrid cars (as well as plug-in hybrids) as step 4 or 5 in a 12-step recovery program to reduce our societal oil addiction. Hybrids are not the silver bullet—it will take more than just improving our mileage to really bring down our oil consumption. But the transportation sector accounts for two-thirds of all the oil we use. There are over 200 million conventional gasoline vehicles (that are registered) on the road today, and about 8 percent are replaced for newer vehicles every year. If we figure out our transportation issue, we'll go a long way toward addressing our energy challenge.

Ultimately, climate change concerns may make burning fossil fuels on a broad basis simply too damaging or costly to continue. In the meantime, hybrids and EVs represent an important step toward weaning the most fossil-fuel dependent sector of our economy off of petroleum. Whether through government mandate or the price pressure of the free market—or a combination of both—cars as we know them are in for a major overhaul, and fuel economy is at the top of the to-do list.

Fuel Cells and Hydrogen ## 12

Hydrogen

Hydrogen is the most plentiful of elements in the universe. It comprises 75 percent of all matter that exists. It is found in great abundance everywhere from distant stars and galaxies to our very own planet earth. In water, hydrogen helps to sustain life. In other forms, such as tritium, a radioactive variant of hydrogen, it can be used to trigger a nuclear fusion reaction in hydrogen bombs.

Lately, this versatile gas has also been used as a fuel for vehicles. Hydrogen is a potential alternative to fossil fuel that has shown promise in recent years, and major car manufacturers are researching hydrogen's effectiveness in powering vehicles.

But most hydrogen cars are really just electric cars masquerading behind a fancy power source. They have electric motors and all the same components as an electric vehicle, but instead of obtaining electric energy from a battery, they draw power from a fuel cell. Research shows that fuel cell vehicles running on hydrogen would use less energy and emit even less carbon than gasoline hybrid electric vehicles.

So what is a fuel cell? A *fuel cell* is a device that generates electricity by a chemical reaction between hydrogen and oxygen. Hydrogen is the basic fuel, but fuel cells also require oxygen, which they get from the air.

Fuel cells were used in America's early space program not just to provide electricity but also water. When you combine hydrogen and oxygen in a fuel cell, you get water, which the Apollo astronauts drank on their way to the moon. Back on earth, this lack of harmful waste products or exhaust makes fuel cells especially appealing.

Also appealing is the fact that fuel cells have no moving parts to break, and they operate at incredible efficiency—two to three times that of

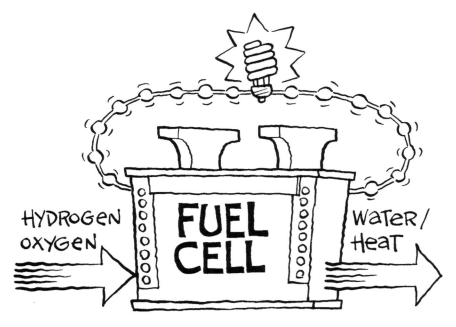

Figure 12.1 Fuel cell. (Doug Jones)

conventional engines. When fully commercialized, fuel cells could trigger a fundamental shift in vehicle technology. Think of what Apple's iPod did to Sony's Walkman.

The U.S. produces over 9 million tons of hydrogen every year. That's enough to fuel more than 34 million cars, but only a tiny fraction of this hydrogen makes its way into automotive applications. In yet another energy-related irony, the vast majority of it is used in petroleum refining and to produce fertilizers.

So why aren't we all driving hydrogen cars? Can't we just make more H_2? The answer is yes, but at great cost. Most hydrogen produced today is derived from natural gas through chemical processing. It can be produced from water, but that process—known as *electrolysis*—requires a substantial amount of energy to separate the H and the O, and there is some question as to whether the end result would be worth it in terms of net energy. (Remember the net energy debate with regard to ethanol in chapter 10?)

For hydrogen-powered vehicles to be broadly adopted by society, they must be cost-competitive with conventional fuels on a per-mile basis.[1]

According to the U.S. Department of Energy, in recent years, the cost of producing hydrogen from natural gas has fallen from $5/gge (or $5 per gallon of gasoline equivalent) in 2003 to $3/gge. Their goal is to reduce this to $2/gge by 2015.

In addition, the cost of making fuel cells must be reduced significantly and the number of hydrogen fueling stations must be increased around the country. The standard the Department of Energy is looking for in hydrogen cars is a range of 300 miles and the ability to refuel in 5 minutes. This would make hydrogen cars competitive with gasoline-powered vehicles, but current hydrogen prototypes don't meet these criteria.

Hope for the Hydrogen Economy

When we talk about electric or hybrid vehicles, we usually just talk about the vehicles themselves. So why do we talk about hydrogen vehicles needing a whole "economy" to back them up? In a word, infrastructure.

We don't currently have a distribution system for hydrogen like we do for oil (pipelines, tankers, trucks) or electricity (generators, transmission lines, substations). In order for hydrogen to become a viable alternative to these two, a new infrastructure to support it will have to be built. The U.S. currently has about 100 hydrogen refueling stations.[2] By comparison, our country has roughly 180,000 gasoline stations, and electric vehicles can be recharged anywhere you can find a wall socket. Clearly, hydrogen has a long way to go just to match the convenience of refueling with gas or electricity.

This lack of a hydrogen infrastructure has created a chicken-and-egg problem much like the situation natural-gas vehicles face, only worse. Consumers are not likely to buy hydrogen-powered cars until hydrogen fuel is widely available at a reasonable price, and fuel suppliers will not spend money to build hydrogen refilling stations unless there are enough cars on the road to use them. Investing in such infrastructure could potentially have huge payoffs for society, but as always, is it worth the cost?

Take railroads, for example. The railroads were a dramatic improvement over horse-drawn wagons or carts. The early steam engine tramway locomotives could haul loads of several tons including several hundred men and women for miles on end, a feat that horses simply could not accomplish.

The invention of the steam engine and the steam locomotive were no doubt critical for the success of the railroads, but it took a lot more than that for railroads to become commonplace in society. In fact, it took land grants, government bonds, and a standard track gauge to allow different trains to share the same path, among other challenges. It even took an act of Congress—the Pacific Railway Act of 1862 was approved and signed into law by President Abraham Lincoln.

We probably won't have to turn over 10 percent of America's land mass to hydrogen producers, but it's a safe bet that a substantial amount of government intervention will be required if a robust hydrogen economy is to take hold in the U.S.

How Do You Make Hydrogen?

Hydrogen is the simplest element known. It is also the most plentiful gas in the universe. Stars are made primarily of hydrogen. So how is it captured and used to drive vehicles?

One approach is to make hydrogen from water—sounds like science fiction, doesn't it?[3] It's up there with transporters to beam us from one place to another and replicators to instantly synthesize a cup of hot Earl Grey. But this is not *Star Trek*, and hydrogen can indeed be derived from water.

Key Concept: Energy Carrier vs. Energy Source

An energy source is a substance that is simply that, a source of energy. Fruits and vegetables are sources of energy for our bodies. Coal, oil, and uranium are other sources of energy. An energy carrier, on the other hand, is something that can store energy or move energy from one place to another.

Electricity is the most common energy carrier. We use it to transport energy from power plants to homes and businesses. Like electricity, hydrogen is an energy carrier and must be produced from another substance that contains the energy to begin with.

Energy stored or transported in the form of hydrogen can either be burned in an internal combustion engine or merged with oxygen in a fuel cell. Both methods deliver energy to turn the wheels. The combustion option releases energy the same way a gasoline engine does. Fuel cells release the energy as electricity, which can then be used to drive an electric motor.

BMW is working on a car that burns hydrogen in an internal combustion engine.[4] The car reportedly will include a dual-mode drive that allows it to run on both gasoline as well as liquid hydrogen, which the driver can fill up at a station much like filling up with gasoline. Much more common, however, is the idea of using a hydrogen fuel cell to run an electric car. That's what most people mean when they refer to "hydrogen-powered cars."

The basic process is known as electrolysis and it's nothing new. In fact, it has been understood for over 100 years. Simply combine water with an electrolyte and then run some electric current through the solution. Voila! The water molecules split into hydrogen and oxygen gases.

So what is the catch? Once again, the culprit is cost. Hydrogen made from electrolysis has been hampered by the very high cost of the metals needed to serve as the electrolyte in the process (e.g., platinum). It also requires a good deal of energy, which of course isn't free either. So until the cost of this process comes down, hydrogen will likely continue to be produced the old-fashioned way. That's why currently over 96 percent of all the hydrogen produced is refined from fossil fuels (natural gas, oil and coal). A mere 4 percent is made by electrolysis.

The Road Ahead   **13**

The days of the gasoline-powered internal combustion engine are numbered. The only question is how many it actually has left. Eventually, our supply of oil will become so expensive that it will cease to be useful as a motor fuel for the masses. It's likely too that the environmental cost of using fossil fuels to power our cars will become so burdensome that we will be forced to look to alternatives.

At this very moment, the automobile is undergoing its greatest metamorphosis in a century. So, where are we headed? When do we get our Jetsons space ships?

The Role of Government: Today and in the Future

Government, both at the federal and state level, plays a huge role in energy as well as transportation. Whatever direction we end up taking, it's likely that our course will be determined by policy decisions made in Congress and in state houses and legislative offices across the country.

The federal government, for example, has been making energy and transportation policy for many decades now. There's a transportation bill, it seems, every year or so, with billions of dollars for a variety of projects—highways, mass transit, air, and rail transport, and so on. And there are enough taxes and tax credits to fill a book. In this section, however, we will limit our discussion to the provisions of one particular act.

At the federal level, the Energy Independence and Security Act of 2007 (EISA) set down a number of policies for transportation.[1] Here's a quick breakdown of some EISA provisions related to transportation:[2]

Fuel economy: EISA mandates an increase in automobile fuel-economy requirements (CAFE standards) to a fleet-wide average of 35 mpg by 2020, an increase of 40 percent over current requirements.

Renewable fuels: EISA requires that half of all new cars manufactured by 2015 be capable of running on 85 percent ethanol or on biodiesel fuels (flex-fuel vehicles). The law also mandates the production of 36 billion gallons of ethanol per year by 2022, which is seven times more than what was produced in 2006. To meet these requirements, ethanol would be made from corn as well as new cellulosic sources, such as switch grass, wood chips, or agricultural waste.

Price gouging at the pump: EISA makes it a federal crime to charge an "unconscionably excessive" price for oil products, including gasoline. We're not sure what price would be "unconscionably excessive" given the $4/gallon we saw in the spring of 2008, but the law also has a provision granting the federal government the authority to investigate what it sees as potential manipulation of the oil market.

Research funding: EISA establishes financial supports to promote research into fuel-efficient vehicles, hybrids, advanced diesels and battery technologies.

States Take Energy Issues into Their Own Hands

How often have you been stuck at zero mph when the sign clearly states that you are allowed to go 65? More and more of us suffer through rush hour gridlock every year, and every year the average commute distance goes up.

If you happen to be on a freeway in New York State, say the Long Island Expressway and oh-by-the-way just happen to be driving an eco-friendly hybrid, or better yet, an electric car, you can zip right past the other sorry looking drivers by slipping into the high occupancy vehicle (HOV) lane. It doesn't matter that you are driving the car all by yourself.

In early 2006, New York Governor George Pataki initiated a program known as "Clean Pass-HOV" that allows single drivers to use the Long Island Expressway High Occupancy Vehicle lanes provided they're cruising in a hybrid or EV.[3] The program was part of the governor's energy independence plan, first unveiled in 2006, and is similar to policies in other states such as California.

In addition to the federal government, state governments across the country are taking several steps to raise fuel economy standards in vehicles, reduce our oil dependence, and fight global warming. The states' role ranges from tax credits for vehicles that meet the emission standards (electric cars, hybrids, and clean fuel cars) to programs that mandate a certain percentage of new vehicles on the road be zero emissions vehicles. Some counties and townships are even offering free parking for hybrids.

When it comes to trailblazing vehicle legislation, a few states stand out from the pack. California's large population, robust and diversified economy, and long history of trendsetting environmental policies make the state a national leader in the development of cleaner vehicles. Policies developed there are often adopted by other states and serve as a model for federal regulations.

For example, the hybrid vehicles now popular nationwide and the hydrogen-powered vehicles being tested today are a direct result of California's Zero Emission Vehicle program. The state has also launched a Hydrogen Highways initiative focused on ensuring that hydrogen fuel will be available as automakers bring zero-emission cars to market.

California was the first state—in the U.S. or the world—to enact standards aimed at reducing global warming emissions from vehicles. California has acknowledged the fact that roughly one-third of all carbon dioxide emissions in the United States come from transportation sources. The percentage is even higher (40 percent) in the Golden State.

California has also been a national leader in cleaning up the dirty diesel engines used in buses, big rigs, and heavy equipment. After estimating that diesel soot was responsible for 70 percent of state residents' risk of cancer from airborne toxins, California implemented a plan that calls for reducing soot by 85 percent below 2000 levels by 2020.

Look to the Left Coast for the next wave of vehicle regulations, as the EPA has granted the state the ability to regulate CO_2 emissions from cars under the Clean Air Act, and has just recently classified CO_2 as harmful to human health under that legislation.

The Role of Technology: Will Solar-Powered Cars See the Light of Day?

So you might have been wondering while reading about electric cars and fuel cells, if you have an electric motor, why not drive it with power from the sun? Indeed, the concept is perfectly sound. Instead of a battery or fuel cell, you get your juice from solar cells mounted on the car.

Solar-powered vehicles are already here, at least at the prototype stage. Technological innovations and creative designs have produced solar-powered boats, cars, golf-carts, and even motorbikes. The technology to tap into the power of the sun has come a long way since the early days of solar-powered calculators and solar cookers, but alas that old curmudgeon, cost, is getting in the way again.

By far the most important and, yes, most expensive component of the vehicle are the high-efficiency photovoltaic solar cells used in experimental automotive applications. These solar cells are far more efficient than the ones on your calculator, and they cost about 100 times more, too. So, you are unlikely to see solar-powered vehicles sitting on the front parking lot of your local auto dealer any time soon.

On the upside, though, you might not need a solar-powered car to travel in eco-friendly comfort. Given the pace of development with plug-in hybrids and all-electric cars, there may be no need for solar panels on your roof and hood if the electricity you're using came from a solar array on your house.

ENERGY SECURITY **III**

T AKE A FEW ELEMENTS RELATING TO global energy security—the assassination of an oil tycoon, a pilfered nuclear bomb from Kazakhstan, a KGB agent-turned-terrorist who threatens to blow up a new oil pipeline, and a flooded reactor detonating underwater. Mix in a dangerous megalomaniac with plans to rule the world and a liberal dose of stunts, speedboat chases, explosions and shoot-outs and what do you get? The nineteenth edition in the James Bond series, *The World Is Not Enough.*

Not many action movies can claim to play into people's concerns about global energy issues. But this is James Bond, perhaps the single most enduring franchise in the history of film.

In recent years, concerns about energy security have begun popping up in a lot of places—the fact that a Bond movie played on them is just a sign of the times. So what exactly do we mean by "energy security?" Like most issues, it depends on who you are.

For most of us, our first contact with energy security is economic in nature—when fuel prices go up, we feel it, and when they nearly double inside of a year, it can really throw a wrench into how we live. You can expand that concept to a national scale, too. At that level, energy becomes a matter of national security. Dependence on a given fuel (e.g., oil), increasing competition for said fuel from developing nations (e.g., China) and unsavory regimes presiding over large portions of the known supply (e.g., Iran) all make for a good deal of economic "uncertainty" that can really affect how a society functions.

Disruptions in the supply of energy can have disastrous effects, especially when the energy supply in question is electricity. "Energy security" in this context has more to do with the reliability of our energy infrastructure than it does with geopolitics.

Finally, there is energy security in terms of preserving our environment in a condition that will continue to support not just human life but life as we know it, and not just in rich countries like the U.S. but everywhere. Energy security in this context means having the ability to improve one's standard of living using energy in such a way that others may do the same, now and in the future.

In the next few chapters, we'll examine the issue of energy security from a few vantage points. This is by no means an exhaustive discussion, but it should help to illustrate the complexity and the gravity of this part of the energy landscape.

Security, Reliability, Diversity, and Independence

14

S UCCESSFUL POLITICAL CAMPAIGNS OFTEN feature a slogan that resonates with voter segments they intend to persuade leading up to Election Day. In some cases, slogans have made a memorable impression and served to encapsulate what that particular candidate stood for. Bill Clinton's campaign in 1992 captured the mood and attention of the entire nation with a comment first scrawled on a sticky note: "It's the economy, stupid." Similarly, Ronald Reagan's 1980 bid for president asked a simple question that struck at the heart of a similar economic theme: "Are you better off than you were four years ago?"

The title of this chapter sounds rather like a campaign slogan, or maybe just a grab bag of words from a politician's repertoire of election speech sound bites. Elected officials on both sides of the aisle talk frequently about energy security, energy reliability, energy diversity, and the path to energy independence. They may disagree on how we should get there, but they all agree that these are important issues for us to face as a nation.

So what are these issues? Are they related to one another, and if so, how? Why are they important today?

In this chapter, we take on the task of defining these terms and providing you with an overview and a context for why the subsequent chapters in this part belong together more than they do apart.

Energy Security: A Historical Perspective

In many ways, attempting to define and explain energy security (and many of the other terms mentioned in this chapter) is to go "where no man has gone before." But like Captain James T. Kirk and his fearless crew, we're on a mission of exploration.

The concept of energy security is at best elusive. This is in part because it has morphed and grown over the past several decades to keep pace with the changing global energy landscape. According to the 2009 IEA World Energy Outlook report,[1] energy security, broadly defined, means adequate, affordable, and reliable supplies of energy.

Historically, this referred mainly to the security of oil supplies. If we could produce enough oil domestically, or purchase enough of it inexpensively from foreign nations, then we were "energy secure." Gasoline would magically flow out of nozzles at gas stations around the country whenever we wanted, and of course, at a price we could all afford.

Definitions of such "oil-centric" energy security have varied around the world depending on whether the country was a net oil producer or a net consumer. Historically, energy security was the concern of oil-consuming nations only. But as we move from an era of states ensuring access (or denying access) to energy resources toward a new era of understanding about the value of an open and stable global energy market, energy security has become the joint responsibility of both consumers and producers alike.

For oil importers like the United States, energy security means the security of oil supply. For oil producers like Saudi Arabia and Qatar on the other hand, energy security means stability of global demand and by extension the stability of prices for the oil they sell. Stability of prices in the open market results in predictable revenues to support their rapidly growing local economies (seen any photos of Dubai lately?).

Energy Security: The Definition Broadens

The focus on oil in international circles has given way to a more holistic view of what constitutes "security" in energy terms. Now it's more of a basket of definitions from which nations pick and choose to suit their own particular circumstances.

Natural gas, for example, has taken on much greater significance and the reliability—or even just availability—of supply has become cause for a couple of recent international spats. The recent Russian-Ukrainian dispute highlighted the importance of natural gas on the geopolitical agenda and how it, too, can play a role in energy security.

In recent years, the paradigm has changed to view the entire integrated energy system as one single interconnected entity.

The supply disruption from hurricanes Katrina and Rita in the Gulf of Mexico is a sobering example of the need for this integrated view. In the wake of the storms, natural gas still flowed in from the Gulf but was then trapped because power lines feeding electricity to the processing plants were severely damaged. The sudden drop in supply drove up prices for gas (and oil), but the disruption itself was caused in part by a lack of electricity, not oil and gas.[2]

Climate change and other environmental concerns may soon also elbow their way into the security discussion if the long-term effects of fossil-based energy are taken to be a threat to security. Aside from the environmental aspects, however, a shift away from oil toward domestically produced alternatives would yield benefits in terms of security associated with reducing dependence on unsavory regimes. Indeed, "energy security" has something for everyone.

Energy Security and Maritime Trade

It may come as a surprise to learn that the U.S. navy routinely escorts oil tankers through the Strait of Hormuz at the mouth of the Persian Gulf. OK, maybe it wasn't a surprise, but even if it was, one look at a map and you'll quickly understand the strategic importance of this particular sea passage when you consider that 20 percent of the world's oil production moves through it every day.[3]

The Strait of Hormuz is a "chokepoint" in energy security parlance, one of a few places on earth where shipments of energy (i.e., oil) flow through a physically small area (21 miles across at its narrowest point). Iran has threatened to seal off the Strait of Hormuz if it is attacked, and the U.S. has countered with a commitment to keep the strait open, no matter what. But the threats to maritime trade in energy extend well beyond one troublesome nation. Piracy, sovereignty disputes, and simple differences in the interpretation of maritime law all present potential problems.

On top of that, the ocean is getting more crowded all the time. Nations formerly regarded as land-loving (e.g., China, India) are now engaged in an expansion of their naval power. Considering that roughly half the world's oil supply is in transit on the high seas at any given moment, it's clear that any discussion of energy security must include maritime security.

Energy Security and National Security

Our nation's energy system is a complex, interconnected web of machinery and equipment. After September 11, 2001, politicians, industry experts,

> ### Key Concept: What Do the Terms
> ### SLOC and UNCLOS Mean?
>
> Sea lines of communication (abbreviated as SLOC) is a term describing the primary maritime routes between different ports, used for trade as well as by naval forces.
>
> UNCLOS is the abbreviation for the United Nations Convention on the Law of the Sea. It defines the rights and responsibilities of nations in their use of world's oceans.

and all of the resources of the newly created Department of Homeland Security increased their attention to the potential for our energy infrastructure to be damaged. This is no small undertaking. The energy system is vast, and has many potential Achilles' heels. Nuclear plants, spent fuel depots, conventional power plants, utility control centers, substations, rail lines (essential for transporting coal among other things), pipelines—the list of potential targets is a long one.[4]

Most of our energy infrastructure is privately owned, but the federal government (not to mention state and local authorities) spends a great deal of resources on protecting it. A lot of that work has to do simply with preventing accidents and keeping systems in good working order, but the issue of security from attack occupies an increasing portion of the protection to-do list.

Energy Security and Economic Security

If you're like us, you probably associate "security" mostly with physical safety. "Economic security" tends to conjure up visions of retirement plans, or at worst the prospect of (not) having enough money to pay the bills. However, if economic stress persists long enough, a lack of funds can have a negative and very real impact on physical health and safety. Navigating that continuum between anarchy and universal wealth is a task policymakers undertake when they consider, for example, how to balance environmental issues with employment objectives. There is also the balance between energy prices and economic development, something especially vital to the booming economies of India and China.

Energy in economic security is increasingly a global problem. As nations grow more and more intertwined—mainly through trade—certain inconveniences are bound to arise. Competition for scarce resources is only the most obvious. Climate change, and more specifically the measures required to mitigate it, carry sweeping economic implications, most of which center on energy use. Are we really going to tell the Chinese they can't all drive cars or consume electricity like we do?

Key Concept: Green Jobs

Green job creation is an important way of linking the local economy with energy security. So what is a *green job*? Green jobs are ones that somehow help our economy transition from a fossil fuel based economy to one that relies more on renewable and alternative energy sources. They might include, for example, manufacturing, selling, delivering, installing, servicing, and supporting clean energy technologies like solar panel installations, wind turbines, and geothermal pumps.

In addition, green jobs are those that support energy conservation and efficiency measures such as LEED certification of buildings, manufacturing and installing energy efficient systems at homes and offices, performing energy audits, and so on.

Green jobs are also a growing segment in the transportation sector. There are many jobs tied to development and delivery of consumer and commercial vehicles that wean us away from our petroleum dependence. These include jobs that are related to biofuels, hybrid vehicles, battery manufacturing, and fuel cells, to name just a few.

Energy Security and Energy Policy

Government policy can play a huge role in energy security. As we discussed in Volume 1, governments the world over have developed policies to protect against failures in the energy supply system that arise from weakness in market mechanisms or issues that cannot be handled by the market alone. Through various incentives, tax breaks, and often simply providing clarity on policy, government can aid the steady supply of energy. This is especially visible in how government programs (i.e., subsidies) have been used to encourage the growth of renewable energy sources.

There are numerous factors that impact energy security in the short term and long term, and many of them are truly out of the control of any government, including ours. Natural disasters come to mind. Price spikes in global markets might seem like a disaster too, but alas they are made by people. The good news there is that policy can help to stabilize prices, at least in the short term.

Some other areas where government policy can play a role include the following:

- Ensuring adequate investment in production, processing, transportation, and storage capacity of various energy commodities to meet projected needs.
- Promoting more efficient energy use.
- Encouraging diversity in fuel mix, across geographic sources, and between types of supply, transportation routes and market participants.
- Supporting market transparency to help suppliers and consumers make economically efficient investment and trading decisions.
- Providing a clear regulatory framework that lets producers know what the rules of the game are with regard to environmental and commercial issues.
- Opening new deposits of oil and gas to production.
- Streamlining the siting process for transmission lines, LNG terminals and other energy infrastructure.

Energy policy on the international level also plays a major role in our energy security. Obviously the policies of producer nations will have an impact, but we're talking here about the U.S. contribution to international policy. For example, the International Energy Agency—of which the U.S. is a member nation—plays a vital role in managing energy crises like the disruption of supply that followed hurricanes Katrina and Rita in 2005. Similarly, the Extractive Industries Transparency Initiative[5] (EITI) seeks to ensure that people living in resource-rich nations reap the benefits of their country's energy exports and avoid the "resource curse" that has afflicted many such locales.

Energy Security and Foreign Policy

OK, we promise we're not going to refer to the Arab oil embargo again. Really, we're not. There are plenty of other examples of how energy security finds its way into foreign policy. In fact, the two are linked perhaps more closely now than ever before.

On New Year's Day 2006, Russia cut off natural gas supplies to Ukraine over a payment dispute. The thing about this particular move was that it also had the effect of cutting off supplies to downstream (or maybe "down-pipe"?) countries. Who lives downstream from Ukraine? Pretty much the rest of Europe, and the disruption in gas supplies eliminated nearly a quarter of all the gas consumed on the continent.

History offers other instances where countries have used their energy resources to further their political agenda in the global arena, but we needn't look further than the morning paper. Iran is making nuclear noises that the West finds particularly troubling. Meanwhile, in 2008, the U.S. made a deal to share nuclear technology with India, a nation that is not a party to the Nuclear Non-Proliferation Treaty (NPT).

The next hotspot for a convergence between energy and foreign policy might be the frigid Arctic. As the hunt for oil and gas moves away from de-pleted fields into new territory, the vast unexplored region under the polar ice becomes ever more enticing. Despite the harsh conditions and fabu-lously high cost of actually setting up shop at the top of the earth, Russia,

---

Key Concept: What Is the "Resource Curse?"

The term "resource curse" refers to an economic paradox in which countries with an abundance of natural resources experience worse economic growth than countries with fewer natural resources. According to John Ghazvinian, a visiting fellow at the University of Pennsylvania and author of the book *Untapped: The Scramble for Africa's Oil*, between 1970 and 1993, countries without oil saw their economies grow four times faster than those of coun-tries with oil. But it's not limited to Africa—the same phenomenon was termed "Dutch Disease" following the decline of the Netherlands' industrial economy after natural gas deposits were discovered there in the 1960s.

So, what accounts for this very counter-intuitive state of affairs?

Well, first of all, oil exports inflate the value of the exporting country's currency making most of their other economic sectors like manufacturing, agriculture, tourism, and so on less competitive on the global market. In extreme cases, it destroys them. Meanwhile, investment is pouring into the booming energy sector instead of those less profitable ones, so a vicious circle is created.

As a result, the country in question becomes heavily dependent on imports for all of these other industries (some even have to import basic food items). They must trade their petrodollars for everything else they consume, exposing the entire national economy to the volatility of the global market for oil.

Then you have good old corruption. A major influx of investment in a single industry creates a certain amount of opportunity for graft or at the very least mismanagement. Sadly, the victims of such diversions of wealth are usually the citizens the government is supposedly there to protect.

Canada, and other nations with something approximating a property claim in Santa's neighborhood are already beginning to jockey for position. The Russians even planted a flag on the ocean floor at the North Pole.

Energy Independence versus Energy Interdependence

Energy independence. It's the bright shiny object that has captured the imagination of most every policy maker and politician for nearly two generations. The idea of being completely self-sufficient is undeniably appealing, but the truth is that the last time the U.S. was really independent, at least in terms of oil, was 1950. Since then, we've come to rely more and more on energy from other places. Today we import more than half the energy we use.

Still, the notion of energy independence has some serious staying power. Every president since Jimmy Carter (with one notable exception) has made a point of promising some form of it. So how realistic is energy independence as a national goal? Not very. We'd be hard pressed to find a single expert who would suggest it's even feasible. What's more instructive is to look at the relative level of dependence we have on other nations for our energy supplies, what exactly we're importing and from whom.

Energy *inter*dependence refers to a more pragmatic goal of systematically reducing our reliance on foreign energy sources but recognizing that we will likely always import some energy from the global market. Energy interdependence also recognizes that we will increasingly be integrated into the global energy market, but we can do so in a manner that is significantly more secure, reliable, robust and less vulnerable than it is today.

The saying "life is a journey, not a destination" might well apply to the path to energy independence. The final destination may or may not be achievable, but that should not deter us from heading down the path. Following are a few ways in which we can reduce our energy dependence while at the same time increasing our overall energy security. If implemented effectively, these initiatives could move our economy toward a cleaner, more sustainable and secure future.

- *Technology.* Technology can make a huge difference in weaning us away from a carbon-based world where most of our energy needs are met from fossil-fuel sources. They range from hybrid vehicles to biofuels to renewables like wind and solar to carbon capture technologies. We have devoted many chapters in our two volumes to describing these technologies in detail.

- *Energy conservation and efficiency.* The Department of Energy projects that U.S. demand for electricity alone will grow 1.1 percent per year over the next few decades. Shaving that figure down through energy efficiency will pay both environmental and economic dividends. The most sustainable megawatt is the one you don't use. As detailed later in this book, such a reduction in demand can come from a variety of sources—homes, commercial buildings, transportation and even programs run by electric utilities themselves.

- *Government policy.* As described in the previous section, there are various initiatives both domestic and international that our government can pursue to help stabilize energy prices and increase our overall energy security. Short-term solutions can range from energy rebates and gas-tax holidays to tapping into the Strategic Petroleum Reserve (SPR) to release more supply into the market. Mid- to long-term solutions might include implementing federal programs to spur innovation in green technologies, providing loan guarantees for construction of new pipelines, increasing fuel efficiency standards, promoting responsible domestic production of oil and gas, and integrating our energy and foreign policy efforts more tightly.

- *Getting more from our existing oil fields.* Bottom line up front: the U.S. has only 3 percent of the world's oil reserves. Having said that, it is important to note that U.S. oil and gas production has and will continue to play an important role in our domestic economy as well as in the global energy marketplace. This means exploring oil in new locations (e.g., the Bakken Shale in Montana/North Dakota is reported to hold 4 billion barrels of oil) as well as leveraging cutting edge technology to extract more juice out of existing wells.

- *Global consumer-producer cooperation.* Just as oil-consuming nations are concerned about the security of energy supply, oil-producing nations are concerned about security of energy demand. The International Energy Forum, established in the early 1990s for the specific purpose of enhancing producer-consumer dialogue, is a good example of a venue where both oil-producing nations and oil-consuming nations come together to discuss common issues impacting the global energy marketplace. The forum is the only major international organization open to both producers and consumers, and its membership includes countries from OPEC, the International Energy Agency and beyond.

- *Leveraging the domestic and global strategic petroleum reserves.* The leveraging of our strategic petroleum reserve (SPR), coordinating reserves with other countries through the IEA, and encouraging emerging economies like China and India to develop their own SPRs will take us further along on the path to energy security.

- *Increased non-OPEC production of oil.* According to the Energy Information Agency, seven of the top fifteen oil-producing nations are not members of OPEC. That list includes the U.S., Norway, and Mexico, all nations whose oil production is now in permanent decline. Reversing that trend might not be possible, but greater diversity in producing nations will likely stabilize energy prices.

- *Transparency and governance of global energy markets.* The energy industry, especially the oil and gas segment, has a long way to go in improving transparency of relevant information for all players involved. Public company data on supply of oil and gas can have accuracy errors of 2 to 4 percent, a substantial amount given shrinking reserves around the world and the fact that excess capacity has for the most part disappeared. Official data is rarely published on a region or a specific reservoir's decline in rate of production. In fact, quite often OPEC member countries view numbers coming out of other member countries with suspicion. There is also strong debate on the amount of remaining oil reserves around the world. In the midst of this are open markets that set the price for energy based on the sparse information they can collect. Commodities like corn, coffee, wheat, soybeans, copper, lead, gold, and silver are all traded in an open and free market allowing forces of supply and demand to determine the price for that commodity. The same goes for crude oil, natural gas, and heating oil, but unfortunately they, too, are not immune from speculation. Energy prices are influenced by fears, forecasts, wars and rumors of war, natural disasters, and even political coups. Most importantly, the lack of adequate transparency contributes to wild fluctuations of energy prices.

Energy Reliability

Oil and natural gas have occupied an overwhelming portion of this chapter's discussion, and that's OK. Undoubtedly, our pursuit of energy security and energy independence has a lot to do with the reliable availability

Key Concept: What Is Energy Diversity?

The United States relies on diverse sources of energy to generate electricity—coal, natural gas, nuclear, hydro, wind, solar, biomass, geothermal, and so on. Our transportation sector relies heavily on fossil fuels like gasoline and diesel. Across all of these sectors, over 85 percent of our energy demands are met by the combustion of fossil fuels.

Energy diversity refers to a conscious effort to shift our economy from one that is heavily dependent on fossil fuels and imports from foreign countries to one that draws on energy from many different sources, both in terms of fuel type (wind, water, sun, gas) and geography (domestic, imported).

of these fossil fuels. But this chapter will not be complete without a short overview of energy reliability, which as it happens is the focus of our next chapter.

We'll save the detail, but in a nutshell, reliability is really about our capacity for *not* thinking about our energy supply. If the system is working properly and everyone does their job, the light comes on when you flip the switch. The gasoline comes out of the pump. The stove ignites when you turn the knob.

In the case of electricity and natural gas, this minor miracle that we take for granted every day is the result of a complex network of transmission and distribution systems that deliver energy from central repositories (power plants, gas reservoirs) to the point of end use. The outlet in the wall or the gas line feeding your oven is located at the end of an enormous but very efficient supply chain.

Energy reliability focuses on the reliable operation of that supply chain for all energy commodities in a manner that is most efficient and cost effective for the consumer. In the case of electricity, reliable operation means preventing a massive blackout and in the case of imported oil, it may mean the reliable transit of ships and tankers through politically sensitive chokepoints around to world to get to their destination. In any case, the goal is the same: keep the energy flowing and the customer happy.

Our Aging Power Grid　　　　**15**

E ARTH HOUR IS A GLOBAL EVENT that began in Australia in 2007.[1] The idea is simple: for one hour, homes and businesses effectively go "off the grid," turning off their lights as a way to raise awareness of climate change. Over two million Aussies took part in the first Earth Hour, and in 2008 fifty million people joined the now global demonstration, including some of the world's most recognizable landmarks from the Golden Gate Bridge to the giant electric billboard in Times Square. For one hour, all was dark.

In many parts of the world today, power outages are a daily routine, but even in advanced industrial nations like the U.S. sometimes we experience an impromptu "Earth Hour" or even "Earth Hours." Elevators stop, traffic lights go dark, and ice cream melts. These minor inconveniences are typically accompanied by some considerably more serious consequences.

In this chapter, we'll provide an overview of power outages, what causes them, how we keep them from happening, and why our power grid isn't as robust as it might appear. You might think this would have been covered in Volume 1 in the chapter on electricity, but we've put this particular topic here because the vulnerability of the power grid is a prime example of how energy security touches our everyday lives.

Mother Nature's Fury

Bruce Willis's character John McClane from the *Die Hard* film series hasn't enjoyed nearly the longevity of 007, but in the most recent installment he, too, fights a foe bent on disrupting the energy infrastructure. In *Live Free or Die Hard*, an ex-FBI cyber security expert named Thomas Gabriel sabotages the nation's network of traffic signals, rail transport, and air traffic control, all with the help of a few clever computer hackers with an Internet connection.

Key Concept: Who Are Electric System Operators?

Our national electric system is made up of a complex interconnected network of power plants, transmission lines, distribution lines, and our homes and offices, which make up the electric load. Electricity is generated in power generators and it is transported to our homes and buildings through the interconnected highways of the transmission and distribution system.

The physical electric system is undoubtedly the largest man-made machine in the world, but this giant system needs to be managed closely to ensure its proper functioning. Since electricity travels at the speed of light, it must be generated at exactly the same instant it is consumed. The supply (all of the energy that can be produced from the various coal, natural gas, nuclear, wind, solar, geothermal, biomass, and hydropower plants) and demand (all of the electricity needed to power our homes, offices, buildings, and factories) must be kept in balance on a nanosecond-by-nanosecond basis.

Figure 15.1   The control room at the California Independent System Operator. (Don Satterlee)

When supply and demand are in balance, the voltage levels at the power sockets in your home are normal and the system frequency stays at 60 hertz. Digital clocks and computers run properly, and everyone is happy. But if supply and demand go out of balance, it could result in serious consequences for everyone on the grid. Voltage levels could fluctuate; system frequencies could drop below 60 Hz or rise higher, thereby damaging electronic equipment. In the worst-case scenarios, portions of the electric system could malfunction leading to a power outage. There is a certain amount of slack in the system, but it still requires constant attention.

System operators are highly trained people working for the electric utility whose job it is to manage this complex system. It is their daily task to constantly adjust output from the various power plants to ensure that everyone's electricity needs are being met. They work in a control room, which in many cases looks quite similar to NASA's facilities of the same name used to command the first manned mission to the moon. But unlike tourists who can visit the NASA Space Center by paying a nominal entrance fee, if you want to visit your local utility's operation control centers, don't expect to see a big sign outside that says "Power Grid Control Center." For security reasons, these facilities are located out of public view, often in nondescript buildings without any outward indication of what goes on inside.

John and a benevolent hacker named Matt Farrell (played by the guy in the Mac commercials) believe that the power grid will be the next system to be attacked. In one scene, the cyber terrorists even redirect natural gas through the pipeline system to blow up a power station. Sure enough, McClane and hacker Farrell escape just in time.

So, how much of this is possible? To what extent are our power grids vulnerable to cyber attacks?

Cyber attacks on power grids have indeed been reported by the CIA. The hackers disrupt power equipment that is controlled remotely by computer systems used by the utilities, but to put things in perspective, such occurrences account for a fraction of a percent of the power outages in the United States. Utilities around the country have implemented multiple layers of security around the systems that control the grid, and even gaining access to them would be extremely difficult.

Of course, there is always room for human error, either while using the computer system or while out in the field working with the equipment that makes up our electric system. But again, human error accounts for barely 1 percent of all power outages in the United States.

So what causes most of the power outages? Who is the main culprit in plunging city blocks into darkness? Well, the question isn't really "who" but "what" since over 70 percent of all power outages are caused by natural events.[2] More than two-thirds of all power outages in the U.S. are caused by weather-related incidents like strong winds, lightning strikes, floods, and so on.

Animals contacting wires (recall the case of the unlucky squirrel on the transmission line that we discussed in Volume 1) and auto accidents damaging poles or other equipment account for about 15 percent of all power outages. About 4 percent of outages are planned and carried out by electric utility personnel to maintain and fix equipment on the grid. A good 11 percent of the causes of power outages are still unknown.[3]

Blackouts, Brownouts, and Power Interruptions

Power outages come in a variety of shapes and sizes. They vary by season and can last from less than a second to several days. The "blip" variety happens more frequently, and they are often just part of the grid taking care of itself. For example, if a tree falls on a power line on your street, your power might go out but people elsewhere on the system will see only a quick flicker, if they even notice anything at all. That's because a circuit breaker did what it was supposed to do and interrupted the flow of electricity in order to prevent further damage to equipment on the grid, not to mention your house's electrical system.

Minor disturbances like this won't affect things like lamps or refrigerators, but they might be enough to disrupt sensitive electronics. That's what would typically cause you to come home to find the clock on your VCR blinking. Either that or you just never bothered to program it in the first place and just got used to it after a while.

Full-on blackouts are a much different story, and they are far less common. To use an analogy from the animal kingdom, if the blips are like tiny hummingbirds (more like the flapping of the wings of a hummingbird) then large-scale blackouts are akin to the giant flying reptile pterosaur Quetzalcoatlus that lived over 65 million years ago and had a wingspan of over 30 feet.

Blackouts begin very much like a regular power outage—when a tree falls on a power line, lightning strikes a transmission tower, or when a generator in a power plant suddenly stops functioning because of an equipment malfunction. But there is one key difference: blackouts cascade, which means they grow and spread. This can happen in a matter of seconds. Much like an avalanche that can build up tremendous momentum and move across a large area very fast, a major blackout can spread to many parts of our national power grid, blanketing several states in darkness.

Although rare, cascading blackouts do happen. In fact, that is exactly what happened at approximately 4:15 in the afternoon on August 14, 2003, when over 50 million people in the Midwest, Northeast, and Ontario were plunged into darkness.[4] The U.S.–Canada Power System Outage Task Force identified the initial cause as a transmission line sagging into a tree, but a series of events following that one created an outage that shut down more than 500 generators at 265 power plants across the region.

Blackouts like the one in 2003 are very rare, but ironically it is perhaps their rarity that in part makes a major outage so notable.

Now what about a brownout? Well, this is not a break in the delivery of power, but rather a kind of dialing back. A *brownout* is a condition where the voltage on the power line drops below normal levels (typically a reduction of 8 to 12 percent is considered brownout territory). The situation can persist for less than a minute or could go on for the better part of a day, and can cause lights to flicker and dim, for example.

So what causes these brownouts? If the electric system is already running close to maximum output due to a prior unforeseen event and supply and demand get out of balance in certain pockets, voltages and fluctuations in system frequency will start to appear in different parts of the distribution system. Sometimes homes at the end of very long distribution lines can experience voltage "sags" because of the physics of electrical losses in the line.

The local electric utility may deliberately lower the voltage to your system in an effort to manage supply and demand during periods of heavy usage. But voltage drops and system frequency deviations beyond a certain point can have damaging effects on many appliances and equipment so these exercises are used sparingly. This is where a surge protector, voltage regulator, or uninterruptible power supply can help (we'll get to those in a moment).

A brownout or a deliberate drop in voltage level at your home is one way for the utility to manage the system when supply and demand get tight. But what happens when they simply don't have enough power generators available to meet the expected load? What happens when there is not enough supply to meet the demand? Utilities resort to a program where they force a blackout in a region as a way of rationing limited power among their customers.

Many utilities offer a schedule, which informs customers as to when, and how long, they will face a power interruption in their neighborhood under such circumstances. This is called a "rolling blackout" or to use an industry term, "load shedding." The blackouts are "rolling" because they roll from section to section of the system, ensuring that everyone shares in the sacrifice of having their power supply turned off for a while.

Developing nations often struggle with rolling blackouts, but industrialized countries are not exempt from them either. In fact, California resorted to rolling blackouts during the energy crisis of 2000–2001. They were not the result of an unforeseen natural event but were planned outages designed to ration electricity because there was insufficient supply to meet everyone's needs.

Uninterruptible Power Supplies

We're not talking about brown delivery vans. "UPS" in this context refers to an uninterruptible power supply, which is an electronic device that continues to supply electric power to any load for a certain period of time during an outage or when the line voltage varies outside the normal limits. UPS systems are typically used in data centers or other facilities where even a momentary disruption in the power supply could cause major problems.

A UPS system for the desktop computers and printers in your home can give you about 15 minutes to save everything you are working on and shut down the computer safely once the power goes out. It is even more useful for the small power blips because the UPS keeps your computer humming along without you even noticing the short loss in power from the grid.

Protective Relays: "To Serve and Protect" the Electric Grid

So how vulnerable is our grid to major disruptions? Normally, the system operators working in the control room have the tools to react immediately to unexpected events. If equipment failure causes a power generator to shut down, these trained personnel can rapidly (in a matter of minutes) bring another generator sitting in reserve online to ensure that supply and demand stay in balance. Think of them as the 911 call center for the power grid. Based on the nature of the given event, they will dispatch the right resources to address the issue (e.g., fire up a gas-fired turbine or instruct a pumped storage hydro plant to begin generating electricity).

But what happens in situations when a few minutes is simply too long? For example, when lightning strikes a transmission tower, a surge of voltage spreads in many directions in a fraction of a second. If not tended to immediately, these voltage spikes can be very damaging to equipment on the grid such as transformers, transmission lines, and even the generators in power plants. This is where the highway patrol of the power grid comes into the picture: protective relays and circuit breakers. Much like the policemen who patrol the interstate, these relays are equipment placed at strategic points on the grid. Their sole job it is to detect potential problems and minimize damage to the grid.

Within a matter of milliseconds, a protective relay can notice an abnormal operation on the transmission grid (such as a wave of voltage from a lightning strike) and inform a circuit breaker to open in order to prevent the problem from cascading further. The protective relay is first to notice an abnormality and brings in the circuit breaker, much like a highway patrolman calling for backup to intercept a speeding vehicle on the road.

No doubt, this operation will cause a disruption of electric service to someone at some level. It may even break up the massive interconnected power grid into smaller independently operating grids, a process known as "islanding," but such actions prevent small problems from becoming large ones. Smart grid technologies aim to make the power grid even less vulnerable to interruptions and blackouts by adding a whole additional layer of "intelligence" to what already exists in order to optimize grid operations while simultaneously bolstering reliability.

Our Aging Power Grid

Speaking of grid vulnerability brings us to another important issue—the age of our transmission grid and its overall frailty.[5] But before we dive into that let's back up a bit and provide a quick overview of our transmission system. Why was it initially created? What was its initial purpose? How is it being used today? What challenges lie ahead?

High voltage power transmission dates back to 1891 when a demonstration was given at the international electricity exhibition in Frankfurt. Most of the U.S. transmission grid as we know it today was constructed between the 1950s and the late 1970s. Over the years, electric utilities connected their transmission grids to serve as backups to each other. Cost sharing and reliability of the power supply were the initial reasons to build and connect neighboring systems. Over time the grid in the United States evolved into three physically separate systems: the Eastern Interconnect, the Western Interconnect, and the Texas Interconnect. Today the U.S. transmission grid consists of more than 200,000 miles of lines operated and owned by 500 companies.

That's rather different from the situation in most other countries where a much smaller number of players own (and run) the transmission system, if not a single state-owned operator.

Investment in new transmission lines dropped significantly after the late 1970s. Since the transmission grid was used primarily for utilities to share reserve generation, and serve as backups for each other, the capacity of the grid was deemed more than adequate. According to the Edison Electric Institute, transmission investment in the United States fell about $115 million each year for twenty-five years, between 1975 and 2000, from $5 billion per year down to $2 billion.

Fun Fact: The New York Blackout of 1965 and the Creation of NERC

The first major U.S. power blackout occurred in New York State in 1965. It was a wake-up call to the power industry. The industry responded to the blackout by creating a voluntary utility-managed reliability organization, the North American Electric Reliability Council (NERC). The blackout also got utilities to increase their cooperation to share reserve energy in case of emergencies.

Grid Modernization: A National Imperative

Although investments in the transmission grid have increased since 2000, there is broad consensus that the grid is not up to the job we've set for it. Built originally to serve existing and future loads, interconnect neighboring utilities, and support reliability, the grid is now also being used to support a larger number of wholesale electricity transactions across regions (we explain wholesale trading in Volume 1).

The role of "trade route" for electricity was never part of the grid's job description prior to the deregulation of the wholesale power industry in the 1990s. Today, power is still "shipped" across large distances under trading arrangements, albeit with a bit more control than during the Enron era. Still, the growth in demand, the proliferation of new power generation including renewable energy sources, and ongoing energy trading all challenge the capabilities of our transmission grid.

The massive Northeast blackout in August of 2003 was more than just a power outage. It also resulted in a loss of public confidence in the grid itself. The general media and the public were left asking when such an event could happen again. The answer was (and is) sobering: another widespread blackout, though very unlikely, could happen again at any time if the same confluence of factors came together at the right time.

Meanwhile, our current transmission grid is also impeding the growth of renewable energy. Sources like wind, solar, and biomass have been hitting their stride in recent years, but will not be able to reach their full potential until our transmission grid is upgraded. One reason is because our current transmission system does not connect the sources of renewable energy, which are often remote, with the population centers that need the energy. The lack of adequate transmission capacity is even putting the brakes on various states' goals to increase their renewable energy portfolios.

Last but not the least there is the issue of transmission bottlenecks or "congestion." This is a problem especially in densely populated areas like the U.S. Northeast where periods of high demand—that proverbial hot summer afternoon—make it very difficult to pull in enough power to serve all those air conditioners while maintaining appropriate reserve margins. It's all but impossible to build a new power plant within a city, so the power to meet our ever-growing demand has to come via transmission lines from somewhere else.

Fun Fact: Did You Know?

- According to the Manhattan Institute,[6] just 20 percent of U.S. GDP was directly dependent on electricity in 1950. Today, it's 60 percent.
- According to the Electric Power Research Institute (EPRI), power outages are responsible for over $46 billion per year in lost productivity.[7]
- According to a 2008 report by The Brattle Group, an energy consulting firm, the electric power industry will need to invest $880 billion from 2010 to 2030 in order to maintain reliable service.[8]

What Is Transmission Congestion?

In the 1993 movie *Falling Down*, Michael Douglas's character William Foster, is an unemployed defense industry engineer who reaches the proverbial end of his rope in Los Angeles traffic on a very hot summer day, and finally just gives up and abandons his vehicle on the freeway.

On June 29, 1956, President Dwight David Eisenhower signed the Federal-Aid Highway Act that funded the interstate system, which has become central to transportation in the United States. It extends over 46,000 miles, but if you count each lane individually the system rolls out 210,000 miles. Fifty years after its inception, the system is overworked and in many places dangerously in need of repair. Segments of it have become so congested that it has long ceased to be a nonstop coast-to-coast highway.

Transmission congestion is no different from the congestion on our interstate system. It occurs when there is insufficient room on the transmission system to ship energy from power plants to areas where the energy is consumed. Typically during peak hours, the transmission "pipes" become choked, not much different from rush hour gridlock (notice the similar terminology, too?).

Going back to our water analogy from the introductory chapters on electricity in Volume 1, you can think of transmission lines as giant pipes that transfer the electric "fluid" from one place to another. These pipes have a physical limit to how much power they can carry, usually measured in megawatts. This limit is known as the "rated capacity" of the transmission line and is determined by its thermal limits (beyond which the line will overheat, sag, and get damaged) as well as voltage limits and other operating constraints.

As utilities start loading their transmission lines to higher levels, the cables heat up, which causes the metal to soften. That in turn makes the lines sag under their own weight, risking contact with trees and other objects. Attempting to operate a transmission system beyond its rated capacity is likely to result in short circuits, electrical fires and serious damage to equipment—not something a system operator would advise doing! Hence, unchecked transmission congestion can also be a potential pre-condition for blackouts.

Obviously, even the most susceptible transmission lines are not congested all the time. Typically, congestion occurs during peak load periods (that hot summer day again), or when demand is increasing rapidly. The trouble from congestion, though, is not limited to physical operating issues.

When congestion occurs in a competitive (deregulated) electricity market, there is a risk of price gouging as energy traders with power to sell are in a position to demand high prices to relieve the congestion. Regulators have the ability to manage this potential for wrongdoing, though, and they monitor market activities to ensure that price gouging does not occur. Still, aside from some operational tweaks and certain smart grid technologies, the only way to alleviate congestion on an ongoing basis is to build more transmission capacity or decrease demand.

Today, system operators in control rooms around the country do their best to optimize the use of the transmission grid given little maneuverability on already crowded lines.

Fun Fact: Who Is the Largest User of Energy in the World?[9]

The U.S. government is the largest single user of energy in the world. The federal government owns and leases thousands of facilities, and energy typically makes up about 50 percent of facility operating costs. It is the largest user of energy in the world, using about 2 percent of all of the electricity and natural gas sold annually in the United States. The largest energy user within the government is the military, accounting for over two-thirds of total federal energy use.

Revamping Our Power Grid: What Needs to Be Done?

According to the American Heart Association, over 7 million Americans have suffered a heart attack in their lifetime. Transmission congestion is not unlike cholesterol buildup that restricts the flow of blood and makes your heart work harder. If left unchecked, the clogged arteries will eventually lead to a heart attack.

Cardiologists often talk about heart disease risk factors. We will do the same in this section. What are the risk factors for our power grid and what needs to be done to reduce them?

Transmission facilities in the U.S. for the most part are owned by electric utilities. However in states that have transitioned to a deregulated electricity market (covered in Volume 1), the utilities have sold their transmission systems to an independent system operator whose sole mission is to reliably operate the transmission grid. (Just to be clear, "transmission" refers to high voltage lines that ship power from power plants to substations. "Distribution" is what your local utility does, though it might own or operate transmission facilities too.)

Building new transmission lines has proven to be difficult in recent years. There are a few main reasons for this.

Public concerns about new transmission lines include land use issues, environmental issues, health issues, and good old-fashioned not-in-my-backyard attitudes (a.k.a. NIMBYism). Unlike solar panels, which increase the value of a home, the typical effect of a nearby transmission line on property values is, well, not so good. Health concerns about the impacts of electromagnetic fields (EMF) make it even harder to get local support. Locating new transmission lines is especially difficult to do because neither the state nor the federal government have the right to force unwilling landowners to part with a portion of their land except under "eminent domain" laws that are almost never used with regard to power lines.

If public opposition prevents us from adding new transmission lines, perhaps technology can offer a temporary fix. FACTS (flexible AC transmission systems) is a term that refers to a family of devices that significantly enhance controllability and increase the capacity of existing transmission lines. Similarly, superconducting materials show great promise in being able to transmit much more electricity than conventional wires and cables, eliminating the need to alter land use to make way for new transmission towers.

Regulatory issues, both at the state and federal level, are another impediment to building new transmission lines. Big transmission upgrades often involve multiple state governments, federal agencies, and regional authorities, all with the ability to stop a project in its tracks. It really gets complicated when you look at who has jurisdiction over the given project.

The part of the grid that is used by the utility to serve its customers (i.e., the distribution system) is regulated at the state level, but the transmission lines often cross state lines, making them the business of the federal government. The feds also regulate wholesale energy trading through the Federal Energy Regulatory Commission. State public utility commissions are also known to implement rules that actually discourage interstate transmission projects. Why? Elected officials in places with cheap power fear that new lines will send that resource out of state, causing rates to go up locally—not so good for voter satisfaction.

But wait, there's more.

In addition to a lack of clarity on the jurisdictional boundaries between states and the federal government, regulatory rules continue to evolve, making them a moving target for utilities looking to invest billions of dollars into the power grid. That in turn creates financial uncertainty regarding how (or how soon) the investment will be recovered, and as we've noted in several other instances, investors don't like uncertainty.

The Energy Policy Act of 2005 sought to streamline the permitting process for power lines at the federal level by synchronizing the review processes of the nine (yes, nine) agencies that might have approval authority over a given transmission project. No word on how that's working out.

The cost to build a new transmission line depends primarily on the voltage level it will use. In today's dollars, the costs vary from about $140,000 per mile for lower voltages like 115 kV all the way to $1,000,000 per mile for ultra high voltage lines of 345 kV and above.[10] The higher voltage lines can ship more power for longer distances than their lower voltage counterparts, but the tradeoff of course is the price.

Utilities will continue to build transmission facilities for which they can recoup their costs by increasing their rates (with the approval of their state PUC of course). Ultimately, then, the cost of upgrading the transmission grid is borne by "ratepayers," a classic industry term for the businesses and homeowners who pay the bills. Not surprisingly, regulators at the state and federal levels are usually reluctant to raise rates, but even if a project moves ahead on the financial front it can still run into years of delays on a variety of other grounds, notably environmental issues.

So why do regulated utilities have to be the only entities upgrading the grid? Why not bring in outside entrepreneurs with their own investors and their own money to build new transmission lines? As a matter of fact, this is starting to happen. It's called "merchant transmission" and a couple of transmission-only companies have emerged in recent years with innovative ways of bringing in investment dollars. Merchant transmission development would also encourage developers to site projects in optimal locations, and promote more efficient use of generation resources. However they still have to cross the same regulatory hurdles as their counterparts in the utility world when it comes to siting.

Conclusion

A modernized grid would reduce the likelihood of a major outage, but it would also mitigate the effects if one were to occur. It would also pave the way for wider use of energy alternatives like wind and solar power, promote efficiency and conservation, and improve our national energy security by making the entire electric system more reliable.

Mitigating public concerns, reducing regulatory hurdles, and providing the right financial incentives are the keys to a successful power grid. Undoubtedly, the public need is pressing.

Nuclear Energy   16

E = mc². We have all come across this famous equation at some point in time in our lives, and the story behind it forms the basis of this chapter.

In spite of everyone's familiarity with the formula, not many people really understand what it means. Albert Einstein's landmark theory says that energy equals mass times the speed of light squared. That's where we come in. By the time you are done reading this chapter, you'll be able to hold your own with nuclear physicists, at least in cocktail party conversation.

Energy from Atoms: Radioactivity, Atomic Bombs, and Electricity

One of Einstein's greatest scientific contributions was the proposition that matter and energy are actually the same thing in many ways. Matter can be turned into energy, and energy into matter, but neither can be created nor destroyed, only changed.

For example, consider an atom of hydrogen. This tiny particle has a mass of 0.000 000 000 000 000 000 000 000 001 672 kg.[1] Pretty insignificant amount, wouldn't you say? What can you really do with that small amount of anything? Not a whole lot.

Einstein tells us something different, though. His equation informs us that this tiny amount of mass can be converted to a lot of energy. To do the conversion, we would simply multiply the mass of the atom by the square of the speed of light (186,000 miles per second) to get 10,000,000,000,000,000 joules. That's equivalent to around 2.7 billion (with a "b") kilowatt-hours, enough to power 245,000 homes for a year and well in excess of the energy used for residential lighting in the entire United States in 2007.

A *joule* in and of itself is not a large unit of energy. One joule is the energy required to lift a small apple one meter straight up or the energy released when you drop a textbook to the floor. Or how about this for a better example: it is one hundredth of the energy a person can receive by drinking a drop of beer.

Just 5 grams of hydrogen can provide about 900 trillion joules. This is equivalent to burning hundreds of thousands of gallons of gasoline. (For any scientists out there wondering how we arrived at this conversion, here it is: One gallon is 3.785 liters. Since gasoline has a density of 0.7 gm/cc, a gallon of gas is about 2.65 kilograms, with about 125 million joules of energy.)

Fun Fact: Did You Know?

Albert Einstein was able to see the power of nuclear energy long before others. Einstein was by no means a hawk when it came to matters of international conflict. To the contrary, he was a peaceful man, and often referred to the wonders of science and nature as being above petty human conflicts.

"Through the release of atomic energy, our generation has brought into the world the most revolutionary force since prehistoric man's discovery of fire," he said in one of his many quotable moments. "This basic force of the universe cannot be fitted into the outmoded concept of narrow nationalisms."

However, Einstein did help to draft a letter to then-president Franklin Roosevelt urging him to fund the development of an atomic bomb before the Nazis or Japan could develop their own. The result was the Manhattan Project, which produced the bombs that the U.S. dropped on Hiroshima and Nagasaki in 1945.

If you want to find out more about Einstein's famous formula, you can visit http://www.aip.org/history/einstein/voice1.htm to hear it from the man himself.

Nuclear Energy 101

Atoms are tiny particles that make up all matter in the universe. The bonds that hold them together are strong, and breaking them releases a tremendous amount of energy. This is the basic concept behind *nuclear energy*.

Nuclear energy can be used to make electricity, but first the energy from the atoms must be released and we humans have figured out how to do that in two ways: nuclear *fusion* and nuclear *fission*.

In nuclear fusion, atoms are combined or fused together to form a larger atom. The process causes bonds to break and re-form in creating the larger atom, and is what goes on inside the sun. This is how the sun produces energy. Nuclear fission works the other way around, splitting atoms apart to form smaller ones, and this is the process used in nuclear power plants to produce electricity.

How Does Nuclear Energy Work?

The children's game of marbles as we know it today gets its name from the practice of making children's toys from broken chips of marble. The Romans played marbles, and likely spread the game across a wide area thanks to the expanse of their empire.

One of the most famous ways of playing the game is to draw a ring on the ground and a starting line just outside the ring. Arrange forty to fifty marbles, all jumbled together, forming a tight circle. Each player's goal is to shoot the cluster of marbles from the starting line out of the ring. The winner is the one who gets the marbles flying farthest all around in different directions.

That is very similar to what happens in nuclear fission. The tightly packed circle of marbles is like an atom's nucleus. The marble being thrown is like a "neutron bullet," which we'll explain in just a bit. The only difference is that the "marbles" in an atom are protons and neutrons and they're arranged in a tight sphere rather than a two-dimensional circle.

When the "bullet" hits the nucleus, a couple of things happen. First, the nucleus splits into two chunks, each having about half the number of protons and neutrons of the original. Second, the splitting of the bonds between the particles releases a huge amount of energy, which appears as heat and radiation.

The process isn't quite that clean, though, and some "free" neutrons are left over and end up flying into still more atoms, continuing the process. This is what's known as a *chain reaction*, and inside of a nuclear reactor it's quite safe. However, if too many neutrons are generated, the reaction can get out of control and a gigantic explosion can occur. That's what happens in an atomic bomb.

To prevent the bomb scenario from occurring inside a nuclear power plant, the extra free neutrons are absorbed by special materials called *control rods* that are inserted into the reaction chamber. By controlling the reaction, it can be harnessed to produce heat—a lot of heat—which is then used to run a steam turbine to produce electrical energy. Uranium 235 is the most commonly used fuel for nuclear fission.

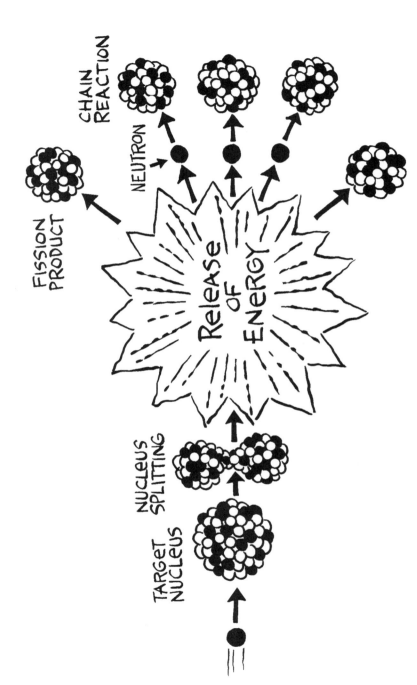

CHAIN REACTION

NEUTRON

FISSION PRODUCT

Release of Energy

NUCLEUS SPLITTING

TARGET NUCLEUS

Figure 16.1 Nuclear fission releases the energy in atomic bonds. (Doug Jones)

Now, back to our "bullet." Recall from high school science that protons are positively charged and electrons are negatively charged, while neutrons have no charge at all. This is key because if you fired, say, a proton at another atom, it wouldn't hit it because the magnetic force of the protons in that atom would repel the similarly charged bullet. Same with electrons, which would be repelled by the target atom's electrons. Neutrons sneak by because they are, well, neutral.

What Is Critical Mass in a Nuclear Reaction?

In his book *The Tipping Point*, author Malcolm Gladwell offers readers a fascinating analysis of how trends occur in society. Everything from diseases to pop culture gets the Gladwell treatment. A tipping point, as he describes it, is the moment that a given trend gathers enough steam to make it unstoppable.

Critical mass refers to a similar idea, wherein there is a particular number of occurrences after which the trend will become self-sustaining.

If a person walking down a city street stops and looks up at the sky, probably nothing will happen. People nearby will go on about their business. But what if five people in a small area all look up at the same time? Would you be inclined to see what they're looking at? We would. What about three people? Maybe, but you get the idea—critical mass is what's required to reach a tipping point.

In nuclear fission, the trend in question is a nuclear reaction, so critical mass refers to the point where it becomes self-sustaining and we don't have to keep firing neutrons at unassuming atoms to keep the reaction going.

This term is now used in many situations outside the world of nuclear energy. Politicians gain a "critical mass" of support, independent movies earn a "critical mass" of viewers, and once a month bicycle advocates in San Francisco take to the streets in a demonstration known as . . . well, you get the picture.

How Do Atomic Bombs Work?

We have all read in history books about the atomic bombs used in World War II. And haven't we all seen at least a couple of big budget Hollywood movies involving nuclear weapons set to be detonated by one villain or another? Recently, atomic bombs have made their way into television too—on the TV show *24*, Jack Bauer struggles to stop a nuclear bomb detonation. In case you're waiting to watch the series on DVD, we won't tell you what happens.

From Stanley Kubrick's 1964 satire *Dr. Strangelove* to the considerably more somber 1980s TV movie *The Day After*, nuclear weapons have occupied a prominent place in our popular culture. In the real world, they continue to fuel apprehension. While nations like the U.S. and Russia strive to whittle down their nuclear arsenals, nations such as India and (perhaps) Iran are actively engaged in weapons development programs.

Nuclear weapons are frightfully powerful, but how do they actually work?

It all begins with *fissile material*, which is simply any material that is capable of sustaining a nuclear fission chain reaction. These substances are unique in that they are composed of atoms that can be split by neutrons in a self-sustaining chain reaction to release enormous amounts of energy. Uranium and plutonium are good examples of fissile materials. Wood and plastic, not so much.

For an atomic bomb to work, it must have enough fissile material so that the chain reaction can get started and reach critical mass. This takes about a millionth of a second. Prior to that, though, the material needs to

Figure 16.2 The Manhattan Project's first atomic explosion, known as the Trinity test, took place on July 16, 1945. (U.S. Department of Energy)

be kept in smaller amounts to prevent the bomb from detonating before it's supposed to. Typically, two pieces of fissile material are kept below critical mass and then smashed together at the moment of detonation to create the chain reaction. However, if they are not combined fast enough or with enough force, a smaller explosion occurs, driving the material apart so that the big bang never happens.

Typically about 11 pounds of nearly pure or "weapons grade" plutonium-239 or about 33 pounds of uranium-235 is needed to achieve critical mass.[2]

Nuclear Fuel

The most important materials used to produce nuclear energy and nuclear weapons are uranium and plutonium, but only their isotopes are directly usable for producing nuclear energy. So what is an *isotope*?

Let's say you have an atom, but it's missing a neutron or has an extra neutron. That type of atom is still the same element since it has the same number of positively charged protons and negatively charged electrons, but it is known as an isotope. They are just a little different from the other atoms of the same element.

Take carbon, for example. The carbon atom we know and love is known as carbon-12, so named because it has six neutrons and six protons in its nucleus. Carbon-13 adds a neutron to make seven, and carbon-14 has two extra neutrons for a total of eight. C-14 is considered an isotope of the element carbon. It is still very much the same element, but just a bit different, for example in that it happens to be radioactive.

The materials most commonly used for nuclear purposes are an isotope of plutonium called plutonium-239 and an isotope of uranium called uranium-235. Uranium-235 occurs in nature, but plutonium-239 must be manufactured.

Approximately 0.7 percent of natural uranium is U-235, the isotope essential for nuclear weapons, and is in fact the only fissile material found in significant quantities in nature. The rest of the world's uranium, around 99.3 percent, is the much more friendly U-238. Uranium mined out of the ground usually contains a mix of isotopes. To convert natural uranium into "weapons grade" material it must be "enriched" to increase its concentration of uranium-235.

Enriching uranium is thankfully both technically difficult and expensive, a fact that has served to limit the number of nations that have thus far developed nuclear weapons.

Nuclear Fusion

A second form of nuclear energy is called *fusion*, which refers to the energy released when atoms are combined or fused together to form another, larger atom.

"Fusion" is also used to describe the combination of things in a wide range of fields. In music, "fusion" is the term assigned to the mixture of rock and jazz. In the culinary world, "fusion cuisine" refers to the blending of different regional traditions. Over at our local star (i.e., the sun), nuclear fusion is what combines hydrogen atoms to create helium, along with an unfathomable amount of light, heat, and other forms of radiation.

Like nuclear fission, fusion can produce large amounts of energy, but it comes with a huge added bonus in that the waste material only remains dangerously radioactive for 50 to 70 years. Compare that to the thousands of years of radioactivity that result from fission and you see why "fusion power" is so attractive. In addition, the raw materials for nuclear fusion, water and silicon, are plentiful and widespread on earth.

Nuclear researchers have been working on nuclear fusion for a long time. If we were able to get it right, fusion power could revolutionize the way we use energy. So far, though, researchers have hit some major obstacles in learning how to control the reaction in a contained space.

Going Nuclear: Atomic Energy's History in a Nutshell

The birth of nuclear energy can be traced all the way back to 1895 when the German physicist Wilhelm Roentgen discovered x-rays. Three years later, Marie Curie discovered the radioactive elements radium and polonium, and by the early twentieth century, Albert Einstein developed his groundbreaking theory about the relationship of mass and energy.

Enrico Fermi, an Italian physicist, first demonstrated nuclear fission in 1934 when his team bombarded uranium with neutrons. In late 1938, two German scientists, Otto Hahn and Fritz Strassman, also demonstrated the workings of nuclear fission using neutron "bullets."

On August 2, 1939, just before the beginning of World War II, Albert Einstein and a group of other scientists told President Roosevelt about a Nazi program aimed at purifying U-235 with the ultimate goal of building an atomic bomb.[3] The U.S. responded with the Manhattan Project, a massive mobilization of science and engineering resources committed to expedite research on and eventual production of an atomic bomb.

From the program's inception in 1939 to the mission over Hiroshima in 1945, the U.S. spent more than $2 billion on the Manhattan Project, which was directed by Robert Oppenheimer and a team of what you could say were the "best and brightest" without the slightest hint of irony. On July 16, 1945, the team got to test their "gadget" and witness the characteristic mushroom cloud of radioactive vapor for the first time. The atomic age was born.

In 1951, an experimental "breeder reactor" (which created more fissile material than it used) in Idaho produced the first usable nuclear power to illuminate four light bulbs. This might sound almost comical today (four light bulbs??), but it was a technological breakthrough eventually leading to the development of commercial nuclear energy.

In 1954, the USSR's Obninsk Nuclear Power Plant became the world's first nuclear power plant to serve users on a grid. It was capable of generating around 5 megawatts of electric power. By the 1970s, American utilities were building dozens of nukes, but the future of nuclear power was dealt a near-mortal blow on March 28, 1979, when a major accident occurred at the Three Mile Island nuclear plant near Harrisburg, Pennsylvania.

No one was hurt, but the nuclear power industry never recovered. A confluence of equipment malfunctions, poor design, and human error led to a partial meltdown of the reactor core inside TMI's number 2 reactor. The release of radioactive material was minute, but public outcry over safety at nuclear plants was anything but. Seven years later, nuclear opponents' worst fears were realized when the Soviet nuclear facility at Chernobyl exploded, exposing millions of people to high concentrations of radioactivity. The disaster has since been linked to many forms of cancer in natives of Eastern Europe and Russia, and has been tagged as responsible for destroying many animals and plants.

The one-two punch of TMI and Chernobyl locked public opinion of nuclear power in the basement for the next two decades. The last nuclear plant to be built in the U.S. began construction in 1977, and no new reactors were ordered after 1979. However, there were at that time several in various stages of construction. It may come as a surprise that the "new" Watts Bar nuclear facility in Tennessee came online in 1996.

More recently, there has been what some industry leaders have termed a "nuclear renaissance." The renewed interest in nuclear power is being driven primarily by climate change—nuclear is the only form of power generation that can produce as much energy in one place as a big coal plant but without the CO_2 and other emissions.

The Energy Policy Act of 2005 included measures to encourage the nuclear industry to start building new nuclear power plants again, but despite the shifting political winds, the U.S. is in fact behind the curve compared to other countries. Nations as diverse as China, Iran and Finland are already building new nuclear power plants, in many cases with American technology.

Nuclear Power Plants vs. Nuclear Weapons: What's the Difference?

Like its primordial cousin, fire, nuclear energy is tremendously powerful and it can be used for good or ill. Leaving aside those connotations for the moment, what exactly is the difference between the "nuclear" in bombs and the "nuclear" in power?

In a nutshell (or bombshell?), it comes down to two things: controllability and enrichment. Nuclear power plants utilize a controlled reaction whereas nuclear bombs (of the fission variety) rely on an uncontrolled reaction. Also, for a bomb to work, the fissile material, whether highly enriched uranium-235 or plutonium-239, must be crushed together and held there in a very precise way. These are some very particular requirements, and they simply don't exist in a nuclear power plant.

In a power plant, the fuel is uranium and is still enriched, but as noted earlier is actually a different isotope (U–238) from the bomb-making variety (U–235). A nuclear power plant also has an array of means to control the chain reaction so as not to allow it to get out of hand. The main way of doing this is with "control rods" made of a material such as boron that can absorb a lot of stray neutrons. By introducing this material into the reaction, excess free neutrons are captured by the control rods before they can strike another atom and produce even more free neutrons.

The uranium fuel also comes in rods, which are in turn made up of pellets less than an inch long. Each one packs a wallop, though—energy equivalent to 150 gallons of crude oil. A bundle of these fuel rods are known as an *assembly*, and one refueling will allow a typical nuclear reactor to run for several months nonstop.

The fission process also creates radioactive waste, and as you probably know, there isn't an easy way to deal with it. Mostly it's stored on-site at nuclear power plants in specially designed containers that are submerged in pools of water. The waste can also be reprocessed to be used again, but that is more expensive and also produces plutonium that could be used to make weapons. The U.S. has a standing policy of not reprocessing nuclear waste.

By contrast, France, which generates over three quarters of its electricity from nuclear power plants, uses reprocessing to manage its nuclear waste.

What Is Radioactivity?

Radioactive elements such as uranium, thorium, and plutonium break down fairly readily. The energy that is released in the process is made up of small, fast-moving particles and high-energy waves. These particles and waves are, of course, invisible but they are what we refer to as "radiation."

So are all kinds of radioactivity bad? Not necessarily, especially if they are used in very small doses. For example, radioactivity is used to sterilize food and medical instruments. A low dose of radiation kills bacteria and other micro-critters that could otherwise cause infection.

Here's another example of radioactivity right from your living room. Smoke detectors make use of the isotope americium-241. This isotope emits alpha particles that are used to ionize air. The smoke detector measures changes in the ionized air by running an electric current through it. If the resistance in the current increases (i.e., due to smoke in the air), the alarm goes off.

To distinguish between radiation that can harm the human body and radiation that cannot, they are separated into two categorizes, ionizing radiation and nonionizing radiation.

Ionizing radiation removes electrons from atoms, leaving behind electrically charged particles called ions. Forms of radiation like visible light, microwaves (e.g., from your kitchen microwave oven), or radio waves do not have sufficient energy to remove electrons from atoms and hence are called nonionizing radiation. They are harmless.

Ionizing radiation is another story. Being in close proximity to a radioactive object exposes you to a pelting by trillions of very small energy waves. You can't see them, or even feel them, but they will damage the cells in your body. A lower exposure might alter the genes in the cells, producing a cancer that will eventually spread but a great deal of exposure can actually break down cells (and the tissues they make up), killing you much more rapidly.

Types of Nuclear Reactors

If you're in the market for a new television, do you go for plasma or LCD? If you are in the market to buy a computer for home use, do you buy a Windows machine or a Mac? If you happen to be the founders of Google, and it happens to be the year 2005, you are probably torn between a Boeing and an Airbus as your personal jet (they settled on a Boeing 767).

Just as there are different approaches to designing and building flat-panel TVs, airplanes, and automobiles, engineers have developed different types of nuclear power plants. The Coke and Pepsi of nuclear plants, at least the ones used in the United States, are boiling-water reactors (BWRs), and pressurized-water reactors (PWRs).

A BWR uses the water heated by the reactor core to drive a steam turbine, so the radioactive water is actually doing the work. In a PWR, the radioactive water is kept under high pressure to remain as a liquid, albeit a very hot liquid. The hot water then runs through a heat exchanger where non-radioactive water is turned into steam that runs the turbine. The radioactive water then returns to the reactor to be heated again. About 70 percent of the nuclear power plants in the U.S. use PWR technology.

Nuclear Energy and the Environment: Disposal of Radioactive Waste

In terms of harmful emissions like mercury and sulfur oxides, as well as CO_2, nuclear energy is much cleaner than fossil fuel power generation. In fact, when compared to energy from coal, it is very clean. Nuclear energy is America's largest source of clean-air, carbon-free electricity, producing no greenhouse gases or air pollutants, though a small amount of emissions result from processing the uranium that is used in nuclear reactors.

Of course, there is a downside. After fission occurs, the used fuel stays dangerously radioactive for thousands of years and must be stored in extremely secure containers. This nuclear waste is a health risk and problem.[4] Storing it is expensive and if it leaks out into the environment, it can also present a major public health crisis.

Most nuclear waste is "low-level" radioactive waste. It consists of things like protective clothing used by workers that has been contaminated with small amounts of radioactive dust or particles. There are regulations that specifically address the disposal of low-level waste materials so that they don't come in contact with the outside environment.

Spent fuel assemblies, however, are highly radioactive and are stored in specially designed pools that act as both a radiation shield and a cooling mechanism. Some high-level waste is also stored in dry storage containers, but the pools are more common. Older waste, which has had a chance to break down more and is not as radioactive, and less-spent fuel assemblies are also stored in air-cooled facilities that are built of concrete or steel.

The United States Department of Energy has been planning a central repository for the nation's spent nuclear fuel at Yucca Mountain, Nevada.[5] However, the project has run into stiff opposition from Nevada residents, chief among them Senator Harry Reid, who currently also happens to be majority leader and as such controls the legislative calendar in the Senate. While the need for a long-term solution for nuclear waste is obvious, it isn't at all clear today whether Yucca Mountain will ever store a single canister of the stuff.[6]

Public Concerns About Nuclear Power

The cultural history of nuclear energy in the U.S. is epic in its proportions. Billed initially as our energy savior, electricity would become "too cheap to meter." Every modern convenience you could think of, and many more you hadn't, would all soon be making our lives easier. Then the dual issue of what to do with the waste and how to prevent weapons proliferation took the shine off our new energy toy.

By the late 1970s, there was already a substantial antinuclear movement under way in the U.S., not to mention other countries. When Three Mile Island happened, it galvanized the movement. When Chernobyl happened, anti-nuclear sentiment went mainstream.

In the wake of September 11, a new specter appeared on the nuclear horizon. The idea that terrorists could acquire a nuclear weapon was and is frightening in the extreme. No longer is the vast destructive power of these devices controlled by a small group of (presumably) rational state governments. Heck, even some of those are demonstrably less than rational. (The guy with the platform shoes and funny hairdo in North Korea comes to mind.)

Our nuclear plants themselves could be exploited by terrorists to release radioactive material into the environment, even if by the crudest of means. When you add up all of these potential very bad scenarios, you might start to wonder why we even bother. Indeed, many people have asked that very question.

Rock 'n' Roll Hall of Famers Jackson Browne, Bonnie Raitt, and Graham Nash started "Musicians for Safe Energy" in an effort to increase public awareness about the pitfalls of nuclear energy.[7] In addition to organizing "No Nukes" concerts, they pushed Congress to stop underwriting loans that utilities need to build new reactors. Other high-profile figures like Patti Davis, daughter of former president Ronald Reagan, spoke out against nuclear power while her father was in office.

More recently, international attention has focused on states like Pakistan that started off with civil nuclear programs but ended up with weapons programs. That's precisely what has Western leaders so concerned about Iran, and though we made fun of his fashion sense a moment ago, Kim Jong-Il has nuclear ambitions of his own that should not be taken lightly.

The formerly exclusive nuclear "club" is becoming more accessible all the time. It's a widely known fact that the information necessary to build a nuclear bomb is readily available, and has been for years—no Internet required! The only thing stopping you or any of your more industrious friends from building one is the difficulty associated with acquiring and processing the fissile material.

Will We Ever Build Another Nuke?

We believe that dropping partisan A-bombs and sharing strong one-sided opinions on nuclear energy only clouds (or should we say mushroom clouds) the underlying complexity of the issue. That is neither fair to our readers, nor is it the purpose of this book. In line with the rest of the chapters, we will offer both sides of this heated debate and let you reach your own conclusions.

Those opposing nuclear power point to the following deal-breaking issues:

Nuclear waste disposal: Spent fuel is now stored on-site at nuclear plants, and it needs to be kept cool. Loss of coolant is one of the most frightening scenarios because if that were to happen, contamination could spread for hundreds of miles. The result would be immediate (i.e., in weeks or months) deaths numbering in the thousands with many more cancer-related deaths to follow. The U.S. has just over 100 operating nuclear plants and fully half the population lives within 75 miles of one of them.

Radioactivity and impact on human health: Low-level radiation exposure is a health risk in its own right. It has been linked to a long list of problems including birth defects, cancer, and immune disorders, just to name a few. Low-level exposure is a possibility that comes with any nuclear facility, and the effects may be more difficult to detect than a high-profile accident.

Reactor safety: Nobody wants another Three Mile Island, and certainly not another Chernobyl. Enough said.

Terrorism: Critics fear that the countries now seeking nuclear energy capabilities, including Iran and North Korea, are just looking for a back door to nuclear bombs. These same critics also express concern about the possibility of theft in transit if our nation's nuclear waste is to be gathered into a central repository like Yucca Mountain.

Potential for renewable energy: Continued research has brought down the cost of renewable energy substantially. The costs of wind energy, photovoltaics, and ethanol fuel have plummeted from the 1980s to today. In light of this, antinuke advocates ask why we look to nuclear power when it is ridden with so many risks.

Overall financial risk: Nuclear power plants may have low fuel costs, but the costs of constructing the plant (initial capital costs) are very high. Also, since many companies will be entering the market after a thirty-year hiatus, these high costs may be underestimated by a significant amount, making the project uninsurable by private insurance companies. Enter the feds.

The Energy Policy Act of 2005 authorized the federal government to step in and provide risk assurance that amounts to a huge subsidy in the best case, and a liability shield in the worst case, to help make these new nuclear power plants financially viable—a controversial subject in itself. Even with that, the upfront cost of new reactors is so high that private shareholders are reluctant to back new plants.

Additional Resources on the Web

Visit http://www.NukeFree.org for a more complete set of arguments against nuclear power.

The proponents of nuclear energy typically respond with the following arguments:

New and safer technology: The next generation of nuclear reactors, which are being built around the world right now, is based on designs that rely on natural processes like gravity and less on technology that has the potential to fail, like say, a water pump. So far, the Nuclear Regulatory Commission has certified four new reactor designs and is reviewing an additional four.

Procedures and protocols to ensure safety: The nuclear industry points to the fact that there has not been one single injury to a nuclear plant worker in all its 104 power plants and 50 years of service in the United States. Also worth noting is that both the plant design and safety procedures used at Chernobyl were far weaker than what would have been allowed in a U.S. reactor.

Recycling used fuel: A pipe dream in the United States even a few years ago, the goal of recycling spent nuclear fuel has today started to gain credibility thanks to the Global Nuclear Energy Partnership. The program would identify "fuel supplier nations" to enrich uranium and take in spent fuel from "user nations" that use the fuel in nuclear power plants. The idea is to encourage the use of nuclear power by limiting the cycle of fuel processing to a handful of trusted nations and thus minimize the risk of fissile material falling into the hands of terrorists.

Yucca Mountain repository: To address opponents of the central repository, nuclear energy boosters point to another way to store spent nuclear fuel: Dry casks on-site at nuclear power plants, thus preventing the need for transporting them to Nevada.

Carbon-free energy: Nuclear energy is proven large-scale CO_2-free electricity production. While wind and solar energy show a lot of promise, they are relatively small sources of emission-free electricity. Nukes, on the other hand, are terrific at producing reliable, round-the-clock energy with no CO_2 released into the air. The 104 nuclear reactors produce more than

---

Additional Resources on the Web

Visit http://www.nei.org for a more complete set of arguments in favor of a renaissance for nuclear power.

Fun Fact: Did You Know?

- There are approximately 440 operating nuclear power plants around the world and together they provided about 15 percent of the world's electricity. The U.S. accounts for nearly a fourth of all nuclear plants, and gets about 20 percent of its electricity from nuclear energy.[8]
- U.S. nuclear plants (once constructed) produce electricity for 1.76 cents per kilowatt-hour, compared to 2.47 cents for coal and 6.78 cents for natural gas.[9]
- Of the 31 countries that have a civilian nuclear program, some depend more on nuclear power than others.
- In France nuclear power is the most widespread; about 77 percent of the country's electricity comes from nuclear power.[10]
- One ton of uranium produces more energy than is produced by burning several million tons of coal or several million barrels of oil.

70 percent of all carbon-free electricity in the U.S., making them a better contender to replace "King Coal."

Conclusion

Today, the U.S. gets roughly one fifth of its electricity from the 104 nuclear power plants that are still operating across the country. Many of them are reaching the end of their designed lifespan. The question of how (or if) we use nuclear energy will turn on a wide range of considerations, but the biggies are shaping up to be climate change on one hand and terrorism/weapons proliferation on the other. The fact that environmental concerns appear on both sides of the issue gives you some idea about just how difficult this question is.

Depending on whom you talk to, nuclear might be a dangerous relic of the cold war, good for nothing but trouble, or it might be the silver bullet—or at least a big piece of silver buckshot—in our quest for energy sources that don't exacerbate climate change.[11]

While we wrestle with these issues, it's worth pointing out that numerous other countries have already reached their own conclusions about nuclear energy, and many are going all-in. China has plans to build over thirty nuclear plants to reduce its use of coal. Japan is looking to expand its fleet to diversify away from imported fuels (granted, they'd still have to get their uranium from someplace, but presumably that would be easier than maintaining a steady flow of fossil fuels).

Here at home, nine permits are in the approval process for new reactors to be built on existing plant sites. Recent polls have shown a new receptiveness to nuclear too: 63 percent of Americans now favor nuclear energy as one way to provide electricity.

What we ultimately decide to do with (or about) nuclear energy could well be the most important single decision in our energy future. Whatever side of the issue you're on, we hope you'll take the time to get the facts and make an informed decision.

The Geopolitics of Oil **17**

S EVERAL YEARS AGO IN TV LAND, the Simpson family and other residents of the city of Springfield got a taste for the repercussions of addiction to oil. When shaggy-haired, thickly accented Scotsman and groundskeeper Willie tries to bury a class's dead hamster, he strikes oil. This means that the school is now rich. Principal Skinner can't wait to start spending the money from the oil well and so he starts to approve every silly request that comes his way.

Mr. Burns, the evil overlord of the Springfield nuclear power plant, also has designs on acquiring the newfound resource. Principal Skinner rejects him, but when they turn on the oil well for the first time, they find that someone else beat them to it—Mr. Burns! His new slanted well creates a host of problems from injury to the Simpson family pet to the closure of the landmark Moe's Tavern due to toxic fumes.

We're definitely not suggesting that you have to rent *The Simpsons* to get a sense for our dependence on oil and the chaos that can ensue when there is a scarcity of it. (But just in case you are curious, and want to rent it anyway, the episode is from the 1995 season and is fittingly titled, "Who Shot Mr. Burns?")

Of the many sources of energy, oil is undoubtedly the world's most widely used. Rising scarcity of oil can even lead to war. In this chapter, we will discuss the serious relationship between energy and war, and the geopolitics of this very important energy source.

Strategic Importance
We take it as a given today that oil is a vital commodity, and that a steady uninterrupted supply is essential to the global economy. However, oil's dominance is a relatively new feature of the geopolitical landscape.

After all, coal was still the dominant fuel through much of the twentieth century.

War, however, put oil on the strategic map, cementing its status as the fuel of choice. In World War I, Britain converted its ships to oil power while the Germans stuck with coal.[1] It was a risky move because the UK didn't have domestic reserves and so had to rely on imports from Asia. However, the superior speed and operational efficiency of oil-powered battleships proved to be decisive.

In 1944, General George Patton famously ran out of gas—literally—in his push from the beaches of Normandy toward the German homeland. Patton argued ferociously for more supplies, but Eisenhower chose to direct more fuel to General Montgomery's operations along the French coast as he moved to take Antwerp. Patton was furious, and Eisenhower's decision was debated for decades afterward, not least because of the substantial casualties the Allies suffered in the grinding final months before the fall of Berlin. Had Patton been given his precious gasoline, he may well have brought the war to an end much earlier, but his forces may just as well have been decimated by a re-forming German army.

Oil's military value was mirrored in the industrial economy, and by mid-century it was clear that it was the lifeblood of the developed world. Franklin Roosevelt stopped in to visit King Saud on his way home from Yalta in 1945 to build his relationship with the man and strengthen the link between America and Saudi Arabia. Roosevelt told the monarch that his kingdom was "more important to U.S. diplomacy than virtually any other nation," and with good reason. Saudi Arabia's oil fields were the largest yet discovered, and represented long-term security for U.S. energy interests at a time when America's dominance in the world oil market was beginning to wane.

Oil Wars

Later world events would see oil affected by war more than vice versa, at least on a tactical level. The Yom Kippur War of 1973, for example, wasn't so much determined by one side's oil advantage, but it did precipitate the first of the "oil shocks" of the 1970s as prices shot up.[2] Oil itself was now a potent weapon. Even before the 1973 war, the prime minister of Iran, Mohammed Mossadegh, was ousted in a coup orchestrated by American and British intelligence agencies. The primary impetus for the coup: Mossadegh's nationalization of the Iranian oil industry.

Figure 17.1 Oil well fires in Kuwait during the first Gulf War. (U.S. Department of Defense)

The tables turned when the Shah, installed by Western interests, was uninstalled by the 1979 Iranian revolution. Saddam Hussein, still a strapping young dictator at the time, saw an opportunity to seize Iran's oil (among other things) and started the Iran-Iraq War, which would last eight very bloody years and end in a stalemate. Now heavily in the red (financially speaking), Saddam decided to remedy his debt problem by invading Kuwait. He was also a bit peeved by Kuwait's habit of producing above its OPEC-set quota.

The threat Saddam's invasion posed to the flow of oil to the global market was unacceptable to the West. Kuwaiti sovereignty was certainly something to fight for, but it's doubtful a regional war of liberation would have attracted the participation of thirty-four countries had oil not been involved. The irony here was that just a few years earlier the U.S. was buddying up to Iraq as a check on the Soviets' influence in the region via Iran.

The conflicts in the Middle East had a direct impact on oil prices for the obvious reason that so much of the world's oil is found in that part of the world. Other wars, however, have had little or no influence on oil prices. During the wars in Korea and Vietnam, for example, prices remained stable and even declined as the supply of oil was never in danger.

The war in Iraq begun in 2003 was construed by many opponents to it as a "war for oil." While this may not be true in a literal sense, it is at least valid that one of the motivating factors for the invasion was, as in the first Gulf War, the preservation of a stable energy supply to the world economy. It's worth repeating here that Iraq is second only to Saudi Arabia in the oh-so-profitable "conventional oil reserves," much of which have yet to be tapped. Replacing Saddam Hussein's brutal regime with a democratic government amenable to foreign investment would represent a major change in the oil landscape at a time when demand is greater and more diversified than ever.

The Trend Toward Nationalization

As we saw in Volume 1 with the new "Seven Sisters," there is a clear trend toward oil-producing nations taking greater control over their oil industries. The limited supply of all kinds of natural resources, even water, has been brought into sharp focus as more of the developing world moves toward the "developed" end of the economic continuum.

Nationalization, however, has been going on for decades.[3] Mexico seized control of its oil industry in 1938, followed by Iran in 1951 and Venezuela in 1976. Saudi Arabia took a 25 percent ownership stake in Aramco in 1973 and by 1980 owned the company outright.

Oil is a geopolitical trump card, and as the economies of countries like China and India rise to claim their share of the petro-goodies, it becomes that much more valuable. Oil is not simply a source of tax revenue; it is influence in a barrel. It's hard to imagine a government, especially one in a developing country with a desire to raise its population out of poverty, that wouldn't make a move at some point to assume control of its most valuable resource.

One look at the list of big producers and consumers of oil, though, and you can see where interests might collide. China, the U.S., even Europe—none of these big-time consumers has nearly enough oil of its own to supply its growing appetite. So, we go shopping at the world oil market. Some of the vendors are friends of ours (Canada, Mexico, and the UK), and some . . . well, let's just say they appreciate our business but they won't be having us over for tea anytime soon.

All of this international intrigue makes for strange bedfellows, and one dark fact is perhaps more illustrative of this situation than any other: of the 19 hijackers who perpetrated the 9/11 attacks, fully 15 were from Saudi Arabia, in theory one of our strongest allies in the Middle East.

Oil and Party Politics in America

Time for the next installment of "Dating yourself with energy trivia." See how many of these political refrains ring a bell:

- "Drill, baby, drill."
- "It's the economy, stupid."
- "Freezing in the dark."
- "No nukes."

As this highly subjective list illustrates, energy has long been an integral part of the political landscape in the U.S., and the world. We could write an entire book just on oil politics over the last thirty years, but given the very limited time we have with this subject, we'll just stick to a few interesting facts.

Probably one of the most reliable indicators of a given president's popularity—aside from what he or she is actually doing in the job—is the price of gasoline. Cheaper gas, happier constituents. In fact, *Business Week* published an analysis in 2006 that showed a remarkable correlation between rising gas prices and sinking presidential approval numbers since the early 1970s. However, the trend only works one way: the analysis also found that falling gas prices don't necessarily help the Current Occupant's approval rating.

Energy is vital to the economy, and the cost of it has an immediate and broad-based impact. It's no wonder, then, that oil has such a great influence, and the fact that nearly all of us rely on it to get from point A to point B only makes it that much more visible.

Lately, energy has become something of a political football as the irresistible force of economic growth has run up against the immovable object of climate change. Everyone wants clean energy, but no one wants to pay for it. The U.S. enjoys some of the lowest gasoline prices in the developed world, but if the price at the pump goes up substantially, we start hearing calls to drill in the waters of the outer continental shelf, the Arctic National Wildlife Refuge or similar pristine locations. Improving fuel efficiency would go a lot further (and sooner) toward energy independence, but show us an elected official who espouses $4 gas as part of the solution and we'll show you a soon-to-be political consultant.

As we move forward into our "carbon constrained" future, it seems a safe bet that we're going to have to get off of oil, or at least greatly reduce our consumption, if we are to meet the objectives now beings discussed with regard to CO_2 emissions and climate change. At this point, though, the world still runs on oil, and it's going to take a long time to alter that fundamental reality.

ENERGY EFFICIENCY IV

K ERMIT THE FROG, CREATED BY Jim Henson, is one of the most recognizable puppets in history. Reportedly, Henson fashioned the original Kermit out of a green ladies' coat that his mother had thrown away, and made the eyes with two halves of a Ping-Pong ball. Not exactly the ultra-realistic creatures that populate TV and film today, but it was good enough to land Kermit several appearances on the *Tonight Show* and even a guest host spot on CNN's *Larry King Live* in 1994.

Kermit epitomizes the core message of this part of the book: a lot of great things can come out of eliminating waste. We're mainly interested in the waste of energy, but the idea can be applied to many different things. By reducing our energy wastage we can reduce our environmental pollution, cut our dependence on foreign oil, lighten the financial burden of filling up our gas tanks, and even create new jobs.

"Green" (a.k.a. environmentalism) is everywhere, in magazines, television, newspapers, even in movies. Protecting the environment is a topic discussed from the schoolroom to the boardroom, and we're convinced it's here to stay. Of course, a major part of improving our relationship with the planet we live on has to do with where we get our energy and how we use it. We need to be prudent in terms of where and when we use energy, but we also need to improve *how* we use it.

This is where the concepts of energy conservation, energy efficiency and green living come together (and no, those first two aren't the same thing, as we'll see in a moment).

We're going to suggest that you can have it all. We're going to argue that saving energy (and the environment) does not have to be a joyless exercise in self-denial, and that there are steps we can take today with existing technology to greatly improve the way we use energy.

We're not going to provide you with a top 100 list of things to do to give your home or office a green makeover. (There are plenty of books that offer terrific to-do lists.) Our goal is to provide you with a big-picture view of what efficiency and conservation really mean, and how we are already moving toward the goal of sustainable living, another buzzword that we'll try to put into context along the way.

By the time you reach the end of this part, we think you'll agree that, in contrast to Kermit's signature line, it *is* easy being green.

Reducing Energy Wastage **18**

THE UNITED STATES USES MORE energy than any other country, but you probably knew that already. Our nation is large, and we have a lot of big, energy-eating industries here, but we're also eating a lot on an individual basis. A funny thing happens to conventional wisdom, though, when you look into the details. In fact, on a per-capita basis, the U.S. ranks behind Iceland, Canada, and even tiny Luxembourg in terms of energy use.

So what gives?

Energy usage by humanity can be divided into three large categories, each of approximately equal size. One third of the world's energy is consumed in the buildings and homes where we live and work. Another third is consumed in the factories and other industrial facilities that produce the products we use, and the remaining third is used in transportation of various kinds.

Where the statistics get interesting is when you look at what we get out of the energy that we use in terms of economic activity. As a whole, the U.S. uses a lot of energy, but we also get a lot out of it. The measure of how much energy a country needs to produce a dollar's worth of product is called "energy intensity." The U.S. uses 9.8 megajoules of energy to produce one dollar of gross domestic product (GDP). That level puts us just a touch higher than China (9.6 MJ), but more than double the figure for India (4.4 MJ) .[1] Lower energy intensity equates to greater efficiency, but it also means more manual labor is used instead of energy-consuming machines to do the work.

The U.S. Department of Energy categorizes energy use in four broad sectors: transportation, residential, commercial, and industrial. Residential buildings are the homes, apartments and townhouses we live in. Commercial buildings include the office buildings, restaurants, coffee shops, movie theaters, and airports we frequent, but overall we spend more of our time at home.

Simply put, commercial buildings are ones that we visit for some portion of the day, but typically don't live in unless of course you are Tom Hanks in the film *The Terminal*. Tom's character, Viktor Navorski, is forced to live in New York's Kennedy International Airport for over nine months, unable to enter the United States or go back home to the fictional country of Krakozia. He makes do with whatever services and facilities are available to him inside the four walls of the airport to get by on a daily basis. And most importantly, he does so with a smile. In many ways, this is the idea behind energy conservation—using as little energy as possible to meet our daily needs.

In this chapter, we will provide you with a brief overview of concepts of energy conservation and energy efficiency. There are some areas where you, the individual consumer, have more control than you may think. For example, did you know that over half of U.S. energy consumption is in the transportation and residential sectors, both of which are largely controlled by the choices you make as an individual?

The idea that a person would or should think about such things is really quite a recent one. Outside of a few early environmentalists, the concept of limiting one's consumption of anything, including energy, was simply not on the table for most people. Energy was cheap. It was abundant. And we used it accordingly. Then something happened that would change the way we look at energy forever.

The Wake-Up Call

In October of 1973, Arab forces led by Egypt and Syria attacked Israel to begin what would become known as the Yom Kippur War. Within weeks, the conflict moved from arms to economics when OPEC imposed an oil embargo on the U.S. and a 70 percent price hike on oil to Western Europe in retaliation for their support of Israel.[2] Prices at the pump shot up, reaching a preposterous $1.20 per gallon (that's $5.75 in today's dollars), four times the going rate before the crisis began.[3]

The result was an instant recession that rocked much of the Western world. If you're old enough to have had a car or even if you were only old enough to remember sitting in the back seat, you can probably remember the gas lines, the tension, and the overall feeling of helplessness.

It was this singular event that awakened the United States from its energy-eating slumber on a widespread basis. Suddenly, "conservation" actually meant something in dollars and cents. The average American consumer now had a very tangible reason to be interested in how much energy they were using.

Climate change has created a kind of second wake-up call, though without the suddenness of the oil embargo but with far wider implications.

Energy Conservation, Energy Efficiency, and Green Living

Jimmy Carter had been president for just two weeks when he donned what would become perhaps his most remembered piece of clothing, a wool cardigan sweater, in a televised appeal to Americans to reduce their energy use. The message received by those watching was stark, a sort of presidential guilt trip, and it didn't take.

In fact, we're willing to bet your impression of the word "conservation" might even have a certain association with this idea. The point is, our understanding of energy use has come a long way, but we still tend to view it as a zero-sum game—using less means going without, giving something up . . . sacrifice!

Now let's try another word association. When we say "energy efficiency," what comes to mind? Hybrid cars? Compact fluorescent light bulbs? Maybe a new ENERGY STAR fridge to replace the old one in the kitchen? Whatever is in your mind's eye, we're betting it isn't quite so gloomy a picture as what the term "conservation" conjures up.

So, what's the difference between these two things? Aren't they basically the same? Well, yes and no.

When it comes to doing more with less in the energy world, conservation is the "what" and efficiency is the "how." You might *conserve* energy if you recycle your newspapers since it takes less energy to produce paper from recycled newsprint than it does to produce it from trees. But the paper mill that makes the stuff might do so with more or less *efficiency* depending on what kind of equipment they're using and how they run their plant.

See, efficiency is essentially a bang-for-the-buck measure. As we learned in Volume 1, combined cycle power plants generate more electricity from the same amount of fuel used by a conventional plant. They are more efficient because they lose less energy as heat. Similarly, compact fluorescent light bulbs produce the same amount of light using far less electricity compared to incandescent bulbs (and by the way produce far less heat as well).

What we're really talking about is minimizing waste. By *conserving* energy so we use less of it, and by using it more *efficiently* when we do, we'll make our supplies last longer and reduce the negative impact of our consumption on the environment.

In recent years, the terms "green living" and "sustainable living" have emerged in our public consciousness as a broader application of this same idea of minimizing waste. In many ways it is larger than energy, as it refers to an entire lifestyle geared toward using less natural resources. All of these concepts, collectively and individually, are valuable because they benefit our security and economic well being as well as our environment. When energy efficiency is combined with conservation and sustainable living, the benefits are compounded.

Energy Efficiency Begins at Home   **19**

G AS PRICES SHOOTING UP AND monthly electric bills making a re-lentless crawl upward makes for a one-two energy punch that is enough to knock us out cold. In this chapter, we will take the concepts learned in the previous one—energy conservation, energy efficiency, and sustainable living—and bring them into our homes. We will provide you with a list of strategies you can execute to give your home an efficiency overhaul.[1, 2, 3] Call it the *Extreme Makeover: Green Edition.*

By no means do we contend that this list is exhaustive. Further, we are also keenly aware that this is only the beginning. We humans are likely going to have to make some changes in the way we live, but our point here is that there is much we can do right now, today, that is relatively painless. In fact, energy efficiency is by far the most expedient means of reducing energy-related pollution and CO_2 emissions, not to mention our dependence on foreign energy sources. Plus, taking action in some or all of the areas that follow could well save you money in addition to saving the environment.

Conduct an Energy Audit

In the cult TV series *Buffy the Vampire Slayer*, the title character gets some help from a "Watcher," who trains and guides her. Well, this time around, the Watchers' Council has assigned us to train you to slay a vampire of a different kind, energy vampires that drink electric power in the dead of night, when you are fast asleep.

The skinny on home appliances is that they use a certain amount of energy even when you are not actively using them. This "feature" is known as standby mode, but it isn't limited to computers. Televisions and other home electronics use standby mode so they can spring to life a few seconds faster when you turn them on (kind of ironic since, in a sense, they are already on).

So, the first step in this battle is to understand how much electricity you use, what you are using it for and when you are using it. You can accomplish this yourself or get help from your local electric utility. In fact, many utilities offer a free energy audit via their website. Fill out a questionnaire, click the applicable check boxes and you are off to a great start!

You can also hire a professional energy auditor to visit your home for a more thorough scrutiny of your energy consumption patterns: your car, your natural gas consumption, your electricity load, your daily commute, even your plane trips.

Key Concept: What Are Online Energy Audits?

With monthly energy bills burning big holes in our wallets, saving energy is a priority for many. Give Home Energy Saver (http://hes.lbl.gov/) your ZIP code and answer a short questionnaire and the website shows you how much you can save by making your home energy efficient. It also gives you tips to accomplish your goal.

This website is sponsored by the U.S. Department of Energy under the ENERGY STAR program for improving energy efficiency in homes.

Flex Your Power also offers a do-it-yourself home energy audit to help you identify your energy efficiency strategy. Visit http:/www.fypower.org for more details.

And don't forget to connect with your local utility to find out about energy saving tips. They may offer assistance to help you become more energy efficient. You can find a partial list of efficiency incentives in your state from the Database of State Incentives for Renewables and Efficiency at http://www.dsireusa.org/.

Use Technology to Slay Energy Vampires

In utility parlance, the term for appliances that drink electric power even when in standby mode is "phantom load" or "vampiric load."[4] Sounds spooky, and it is, in a way, with all these devices secretly sucking up electricity when they're supposedly turned off. So, how do you slay these invisible phantom loads that lurk in the dark crevices behind your entertainment center? There are a variety of new technologies in the market that help track and eliminate energy wastage.

A number of new gadgets have come on the scene in recent years under the umbrella term of "smart house" products. But you needn't turn your home into a science experiment to kill off a few of your local energy vampires. Below are a few ideas to whet your appetite.

P3 International offers a unique meter called a Kill A Watt™ meter that assesses how efficient a given appliance is.[5] The meter tracks consumption similarly to how the utility does, in kWh, and you can calculate your electrical expenses over a given period of time.

"Smart" power strips allow you to have a kind of master switch for a whole group of devices.[6] When one "master" device is turned off—say, your TV—the rest of the devices are shut down at the same time.

"Smart metering" refers to an advanced type of electric meter outside your home that allows your electric utility to directly and instantaneously read energy consumption information from your home. Many utilities also make this data available online so you can analyze your own energy usage. What this means is that your electric power meter can tell you the cheapest time to run equipment or appliances based on the price of electricity in your region at that moment. For example, you could be saving a bundle simply by running your dishwashing machine and clothes dryer after 8 pm.

Information-giant Google has entered the smart grid world with the Google PowerMeter. This software application links to utility information and shows consumers graphical charts and plots of their power use.

Swap Out Your Energy-Guzzling Appliances for More Efficient Ones

This is the electric equivalent of replacing an SUV with a more efficient hybrid vehicle. But in the case of home appliances, you don't have to give up functionality to realize some serious energy savings. Refrigerators, dishwashers, and washing machines have all improved greatly in the past several years, so if you own one that is more than a decade old, chances are you're paying more than you should for the energy to run them.

The improvements in efficiency have come largely through standards imposed by the federal government, which have been progressively raised over the years. Refrigerators are the poster child: today's best models consume less than half the energy used by an average model from just twelve years ago.

Opt for ENERGY STAR® Appliances

ENERGY STAR might just be the most successful efficiency program the Department of Energy has ever launched. It's straightforward, easy for consumers to use, and makes a very clear case for buying more efficient home appliances.

According to the Department of Energy, if just 10 percent of American homes used all ENERGY STAR products, the resulting drop in carbon emissions would be equal to planting 1.7 million acres of trees.[7] DoE also reports that ENERGY STAR appliances in 2008 cut $19 billion from consumers' energy bills and the equivalent of 29 million cars' worth of GHG emissions.

Use the EnergyGuide Label

If you've purchased a new washer, fridge or other appliance recently, you've seen the EnergyGuide label. It's big, it's yellow, and it's required on all new appliances sold in the U.S. The idea is to allow consumers to compare appliances in terms of energy consumption and operating cost on an apples-to-apples basis.[8]

Try on Jimmy Carter's Sweater

Ok, we didn't really mean that literally. (We don't want you searching for President Carter's cardigan on eBay. We looked—it's not there.) But you can give your monthly electricity bill a trim by being aware of your thermostat setting. Combined with sufficient insulation and well-sealed windows and doors, a programmable thermostat could save you a bundle. (And wearing a sweater won't hurt you either.) According to the Alliance to Save Energy (http://ase.org/), even a one-degree reduction on the thermostat will save you about 3 percent on your heating bill. (Don't let that word "programmable" scare you off; it takes less than a minute to figure out how to set your heating and air conditioning to use less energy at night or when you're away.)

Run Your Appliances During Off-Peak Hours

If you're already on a time-of-use rate with your local utility, then you're probably aware that it makes sense to run appliances like dishwashers and washing machines at night. If you're on a flat rate, you might want to check with your utility about moving to time-of-use. And if they don't offer a TOU rate, ask them why—it makes as much sense for the utility from an operational standpoint as it does for the customer from a price standpoint. Bob and his wife switched a few years ago and now about 85

Key Concept: What Are On-Peak and Off-Peak Hours?

Human activity takes place in cycles. Excluding some of you party animals out there, a few students cramming for their finals, and Vikram's daughter who revels in waking up at 3 am every day, most of us rest at night and leave activity for the daytime. This translates to how we use energy.

Simply put, our total energy consumption varies widely between day and night. If you tracked how much energy your home used every single hour of the day for an entire week, and plotted it on a graph, you'll see that it varies in cycles, up during the day and down during the night.

The hours of highest energy consumption are called the "on-peak" hours, and they typically occur in the late afternoon when air conditioners are working hardest and before people leave their places of work. The energy that has to be provided to consumers during this time period is called *peak load*. The remaining hours of the day are the "off-peak" hours when people are heading back home after work, slowing down their activity and preparing for rest later on at night.

Factories and office buildings are offered different electricity rates based on how much energy they consume during peak hours versus how much they consume during off-peak hours. These rates are known as *time-of-use* or TOU rates. By switching the use of machines and equipment to off-peak hours when possible, customers on these rates can save a lot of money.

Residential electricity customers are now beginning to have more access to TOU rates, but there is an environmental benefit to shifting your energy usage to off-peak periods since it reduces the need for peaking power plants to be built. With more smart meters being installed at homes, we may soon see an increase in the number of residential customers on TOU rates.

percent of the power they use is during off-peak times. We'd be lying if we said the savings were large, but hey—free money is free money.

Don't Flush Money Down the Toilet

Cutting back on your consumption of bottled water will save you money, but in this case we're taking about something different. It is about the money that you flush down your toilet every single day—literally! Did you know that 10 to 20 percent of your state's electricity is used to collect, purify, transport and treat water? [9] Ultra-low-flush toilets use about 20,000 gallons less water annually than older models. You can also purchase low-flow shower heads that will cleanse you of this hidden energy waste.

Reduce your water consumption and you'll reduce your home water heating bill and your state's electricity consumption all at the same time.

Take Advantage of Federal Tax Credits and Energy Rebates

There are several energy efficiency improvements that are eligible for a federal tax credit. They include installation of geothermal heat pumps, solar water heaters, solar panels, small wind energy systems, and fuel cells. Even just choosing a higher efficiency version of a traditional forced hot water heating system (i.e., baseboard heat) can get you enough money in rebates to offset the additional cost.

You need to file IRS Tax Form 5695 with your taxes to claim federal credits. Tax incentives are also available for windows, doors, insulation, HVAC, and non-solar water heaters. You can find a lot more information on federal tax credits for energy efficiency home improvement at Energy Star (http://www.energystar.gov) and at the Alliance to Save Energy (http://ase.org/content/article/detail/2654).

Key Concept: What Is a Federal Tax Credit?

In general, a tax *credit* is more valuable than a tax *deduction* of the same amount. A tax credit reduces the tax you pay, dollar-for-dollar. Tax deductions, such as those for home mortgages and charitable giving, lower your taxable income. So, if you are in the highest 35 percent tax bracket and you get a $1,000 tax deduction, you'll get $350 back at the end of the day. If you get a $1,000 tax credit, every dollar of it goes back to you when you file your return.

A Bright Idea on Light Bulbs

Light bulbs have been around since the early days of electricity and stand as a symbol of innovation and new ideas. From jokes ("How many lawyers/engineers/musicians does it take to screw in a light bulb?") to cartoon thought bubbles, they are a ubiquitous symbol for the usefulness of electricity and good ideas. In recent times, they have also become the symbol for energy efficiency. *Time* magazine felt so strongly about it that they put a light bulb on the cover of their January 2009 issue, a compact fluorescent bulb.

The traditional incandescent bulbs we're all used to are not very energy efficient. Compact fluorescent light bulbs (or CFLs) use 75 percent less energy and can last up to ten times as long as comparable incandescent bulbs.[10] You can expect to save about 700 kWh over the lifetime of each CFL[11, 12]; and depending on where you live, that translates to $30 to $75 worth of electricity per bulb! (As that wide range tells you, it's also worth taking a moment to look at your electric utility bill to see what you're paying per kWh; that one bit of information can help you assess the cost-effectiveness of various efficiency measures discussed here.)

Figure 19.1 CFLs last longer and deliver more of the energy they use as light. (Gina Fesmire)

Fun Fact: Did You Know?

- In 2008, the average household spent a total of over $5,500 on energy. However, only $2,200 of it went to energy costs inside the home. The other $3,300 went to transportation fuel (i.e., gasoline).
- Your heating and cooling costs alone account for about 50 percent of your home's monthly energy bill.
- Energy vampires use 5 percent of our monthly energy consumption and cost consumers more than $3 billion every year.
- Refrigerators are the big energy consumers among home appliances, and can use up to 20 percent of the total electricity in your home. You can make your fridge more efficient by filling it up with stuff (food, or even just jugs of water) since the less air there is, the easier it is to keep everything cold.
- Energy-efficient showerheads and faucet aerators can reduce the amount of water released from a tap by up to 50 percent. Less hot water means less energy to make the water hot.
- Only 25 percent of the total cost of having an air conditioner is in the purchase price of the unit. The other 75 percent goes to the energy used over its lifetime.
- A high-efficiency room air conditioner can cut energy consumption by 20 to 50 percent. Replacing a 10-year-old room air conditioner with an Energy Star model can cut energy bills by an average of $14 a year.
- If every household replaced just one incandescent light bulb with an ENERGY STAR qualified CFL bulb, the energy saved would be enough to light about 3 million homes for a year.
- Around 80 percent of the energy used to wash clothes goes into heating the water. Try washing in cold.

Additional Resources on the Web

Visit the following websites for up to date information on energy efficiency:

- U.S. Department of Energy's Office of Energy Efficiency and Renewable Energy (http://www.eere.energy.gov/kids/smart_home.html) offers games, tips, and facts for kids who want to save energy.
- The Energy Hog campaign (http://www.energyhog.org) at the Alliance to Save Energy focuses on educating the general public on energy efficiency related issues.
- The Alliance to Save Energy (http://ase.org/content/article/detail/2654) has resources to make it easier for homeowners to receive tax credits for energy efficiency home improvements.
- The U.S. Energy Information Administration (http://www.eia.doe.gov/kids/energyfacts/saving/) has tons of energy facts and figures for kids.
- Energy Quest (http://energyquest.ca.gov/about.html) is an educational website developed by the California Energy Commission.
- Flex Your Power (http://www.fypower.org) is an efficiency marketing and outreach campaign run by the State of California.
- The Home Energy Saver project (http://hes.lbl.gov/) is aimed at improving energy efficiency in homes.
- The American Council for an Energy-Efficient Economy (http://www.aceee.org/) is a nonprofit organization dedicated to advancing energy efficiency.
- The Department of Energy has built a website to engage kids in saving energy. There you can download an easy energy action plan that lists ten simple ways to use energy wisely. Visit http://www.eere.energy.gov/kids/pdfs/EnergyActionList.pdf or http://www.loseyourexcuse.gov.

Green Living   20

FOR THOUSANDS OF YEARS BEFORE the Europeans arrived, Native Americans held a special and sacred relationship with the earth and its many natural resources. Many of their customs and ceremonies, still practiced today, show their respect for the land and the life it sustains. They saw earth and its resources as a living network. The idea of "owning" a piece of it was completely foreign. Some tribes ascribed to a "seven generation" principle that is essentially a doctrine stating that the tribe should not do anything that would adversely affect the tribe seven generations into the future. How's that for taking the long view?

Now fast-forward a thousand years to the current day. Modern human inhabitants (i.e., all of us) have quite a different relationship with the earth. The United Nations has concluded that the human population will exceed 9 billion by 2050, but even by the mid-2030s we will reach the point where we need the equivalent of two planets' worth of resources to keep up with our consumption.[1] We could easily find ourselves in global ecological overshoot, which is what scientists call it when your ability to replace or recycle resources is outstripped by your consumption and waste of them.

We may not be telling you anything you didn't already know, at least on a conceptual level. But as the alarm bells have gotten louder, more and more people have begun to adjust the way they live with an eye toward avoiding ecological overshoot (global or otherwise). "Green living" takes many forms, much of it either directly or indirectly related to energy. We'll have a look now at a few aspects of this trend, but again, this is just a sample of a very broad and multi-faceted subject.

Living green involves dealing with energy in a variety of integral ways: energy conservation, reducing your carbon footprint, reducing pollution, reducing energy wastage, and making your home energy efficient. For better or worse, energy-related issues play a huge role in the adoption of a sustainable lifestyle. Next we'll look at how those two topics intersect in several important ways.[2]

Depletion of the Planet's Resources

"Fossil fuels." If ever there was a term with multiple layers of meaning, this is it. Coal, oil, and natural gas are all derived from prehistoric matter, but they are also energy "dinosaurs," fuels from a different time when we weren't constrained by the reality of pollution or the prospect of climate change. Fossil fuels still provide 85 percent of all the energy used in the United States, but they are not renewable—once we use them, they're gone forever.

We'll get to the emissions they produce in a moment, but this last point we can't stress enough. Non-renewable energy sources like oil, coal, and natural gas are finite on earth, and before they actually run out, the supply will begin to be outpaced by demand.

When that happens—even if it's perceived to be happening—we are likely to see a massive price spike or two. The economic implication of this is that as the price of fossil fuels rises (from carbon taxes, cap and trade, supply shortages, peak oil, or all of the above), a larger share of the market (we hope) will be served by alternative energy sources.

The trillion-dollar question is whether that global transition will be smooth (by the time we substitute wind, solar, nuclear, and so on for fossil fuels, the price of fossil fuels will decline, not rise) or choppy (economic shocks and unprecedented levels of social conflict). It's no wonder, then, that remaining fossil fuel supplies have been—and continue to be—one of the main sticking points behind regional and global conflicts. That's also why we spend so much time (a whole other book!) discussing "conventional energy" and have devoted a huge part of this book to alternatives to fossil fuels.

Opting for Renewable Resources

In this section, and elsewhere in this book, we have established that we are indeed going to run out of fossil fuels someday. Experts debate on when that will happen, but all agree that once we consume these fuels, time's up! Economists suggest that price signals will drive us to stop using fossil fuels and switch to viable alternatives long before the last barrel of oil has been extracted from a remote corner of the earth or the last lump of coal stripped out of a mountain.

So what are these alternatives and how "viable" are they really?

Sustainability advocates point to *green energy* as an alternative to conventional energy sources. Wind, solar, geothermal, and marine power make up the main categories, with hydropower being included or excluded depending on your take on the environmental tradeoffs involved with dams.

A growing number of electric utilities offer green power, though it usually costs more than electricity from fossil and nuclear sources. You can buy green power in three ways. Many states allow consumers to choose their provider of electricity, much like we choose long distance phone service. In these states, buying green power is as simple as calling the energy services provider that offers power from renewable energy sources and switching to them. In other states, where your only source of electricity is from a regulated utility, you can ask them if they offer a green power option. This allows you to purchase power from wind farms or solar farms with which the utility has power purchase contracts.

Whether or not these first two options are available to you, though, you can buy the green power attributes separate from electricity, in the form of tradable renewable certificates (also called TRCs). These certificates are created only when a given quantity of electricity is generated from a bona fide renewable energy source somewhere on the electricity grid. By purchasing these tradable renewable certificates, you can be assured that the renewable energy is generated, somewhere, to meet the needs of your home or office.

If this sounds a bit second-hand, you should know that no one really has control over where the electrons flowing into their home actually come from. Unless you're generating your own power, it's impossible to know precisely where or how your power was generated. The point of acquiring green energy is to ensure that an amount of power equivalent to the amount you consume is generated from renewable resources. As more consumers choose these renewable options, utilities and energy service providers will have to expand their renewable portfolios.

Fun Fact: Green Power Programs

The Department of Energy has an entire website dedicated to green power that includes a map that allows you to search for green power programs by state. Visit http://eere.energy.gov/greenpower/.

Global Warming

Oceans of ink and forests' worth of paper have been devoted to this topic, so we're not going to get into the nitty gritty here. (Plus, we have devoted an entire chapter in Volume 1 to it.) It's worth noting, though, that while scientists are notoriously picky about making black-and-white statements, there is overwhelming consensus about two things. First, the earth is indeed getting warmer. That much is agreed upon by everyone, largely because it's measurable—the planet has warmed by about 1°F over the past 100 years. You can look it up. What little controversy there is stems from the question of what is causing this warming to occur, in particular the influence of human activity relative to other natural forces.

Those gases create a kind of one-way insulation. Heat from the sun comes in but it doesn't leave. A certain amount of this heat-trapping is a good thing—if we didn't have an atmospheric blanket, much of the life on planet earth wouldn't be able to survive. Many greenhouse gases occur naturally. Water vapor, for example, is technically a "greenhouse gas." Likewise for methane.

There are also some gases that exist only because humans made them out of other naturally occurring elements, and some of these pack a heck of a greenhouse punch. Sulfur hexafluoride (SF_6) is an inert gas that happens to be very good at insulating electrical components. It allows entire substations to be packed into a fraction of the space, sheltered from the elements. It also happens to be over 20,000 times more potent than CO_2 as a greenhouse gas. Other gases like hydrofluorocarbons (HFCs) and perfluorocarbons (PFCs) also result exclusively from human activity and contribute to global warming.

Then there is good old CO_2, and boy do we release a lot of it. The U.S. alone produces six billion metric tons of the stuff every year—that's one fifth of all the CO_2 from human sources in the world. In the past 150 years, the concentration of CO_2 in the atmosphere has risen by 31 percent.

Energy use is intertwined with CO_2 emissions because it's intertwined with just about everything we do. The average U.S. household is reported to produce around 150 pounds of CO_2 every day. That's a pretty big "carbon footprint," the term used to describe the total CO_2 a given building, family or industrial process produces. Look at all the little things you do over the course of a day and much of it can be tied directly back to energy use. Driving to and from work, cooking dinner, turning on the air conditioner, watching a DVD—it would be easier to list the things that *didn't* involve some kind of energy use.

Buildings and transportation make up the bulk of CO_2 emissions sources at 38 percent and 34 percent of our national total, respectively. It's not hard to see why, when you consider how much energy goes into running an office building or how much fuel we need just to get to the office. One gallon of gasoline produces nearly 20 pounds of CO_2 when burned in a vehicle engine, which means that many of the vehicles on our roads today are burping out more than a pound of CO_2 for every mile they drive.

So, now that we've depressed you, let's have a look at how we might go about reducing the amount of CO_2 and other greenhouse gases we emit.

Fun Fact: Did You Know?

- U.S. households produce more than twice the European average in CO_2 emissions, and almost five times the global average, mostly due to longer driving distances and larger homes.
- A gallon of gasoline burned in your car adds a whopping 20 to 30 pounds of CO_2 to the atmosphere.
- On an average, a kilowatt-hour (kWh) of electricity produced in the U.S. produces 1.5 pounds of CO_2.
- If we converted half of all light bulbs in our country to compact fluorescents, we would reduce CO_2 from lighting by 42.2 million tons a year.
- If we turned off home electronic equipment: computers, printers, cell phone chargers, clock radios, camera battery rechargers, and so on, when not in use, we could cut their CO_2 impact by 8.3 million tons a year.

Better Ways to Live and Work:
Energy Efficient Buildings

As we just learned, buildings account for more than a third of all energy use, but they account for fully 65 percent of electricity use. We admit it's hard to fathom how these stationary objects manage to gobble up so much of our energy resources. Really, where does it all go?

Well, a lot goes out the window, literally. Old single-pane windows, leaky doors, even poorly sealed cable TV hookups all allow the warm/cool air inside your house to escape to the outside, which means you have to make more warm/cool air. Americans also have a penchant for lower-density housing, which means more energy use (and loss) per person versus areas where multi-unit buildings and smaller houses are the norm.

That doesn't mean, though, that we can't greatly reduce our energy use in the homes we already have. Shelter is one of our basic needs, so we're always going to need a place to live. We just need to be smarter about how we build, and there are many different ways that so-called "green building" improves on the typical house of today. The benefits of green building also extend well beyond energy use to include lower ongoing costs, better durability, and improved health for the occupants.

Taking the LEED

Green buildings use non-toxic materials, they conserve non-renewable materials and they seek always to keep their environmental impact to a minimum. Probably the most visible example of the green building movement is the LEED (Leadership in Energy and Environmental Design) system.[3]

LEED standards incorporate everything from energy use to building materials to the actual design of the floor plan, all for the purpose of building structures that will endure over the long haul while having a much smaller impact on the health of people and the environment. As we've seen with energy efficiency, many of the differences between green buildings and traditional structures come with economic benefits as well as environmental ones.

The LEED rating system for green buildings was developed in 1998 to elevate environmental concerns in the building community. Since its introduction, LEED has become the most readily identifiable standard for sustainable building around the world. The system is administered by the U.S. Green Building Council in Washington, DC, a non-profit organization made up of architects, builders and industry leaders. LEED ratings fall into four main categories: bronze, silver, gold, and platinum (we'll let you figure out which one is at the top), according to the level of sustainability achieved across a variety of measures. To find out what separates a gold building from a platinum one, or for more information, visit http://www.usgbc.org.

Better Ways of Getting Around: Sustainable Transportation

The transportation sector accounts for roughly 30 percent of total U.S. greenhouse gas emissions.[4] According to the U.S. Department of Transportation, Americans use vehicles to travel even the shortest distances with

*Figure 20.1 LEED-certified buildings like the PNC Plaza
in Pittsburgh use less energy through a combination
of design and technology. (Wikipedia)*

an incredible 44 percent of all car trips being less than 2 miles. It's no wonder the U.S. Department of Energy expects transportation energy use to increase 48 percent between 2003 and 2025.[5]

Let's face it. We are all guilty of driving our cars all of a mile (or less) to our local gym only to use the treadmill there. Or we hop in our cars to do a quick grocery store run when a bike ride would have been a viable alternative. Eliminating these short car trips could save us over a million gallons of gasoline a year, not to mention tons of CO_2.

You can avoid transportation energy use by following several simple yet effective steps: Shop online instead of making the trip by car; ask your boss if you can telecommute one or two days per week; check your tire pressure frequently and keep your tires fully inflated (remember full tires can improve your fuel economy up to 10 percent) and if you're in the market for a new car, buy a hybrid or the most fuel-efficient vehicle that meets your needs.

This last topic is so important to us that we devoted an entire part of the book to discussing various alternative fuel sources and newer vehicle technologies like hybrids and electric cars.

Additional Resources on the Web

Hopstop.com is a very useful site that educates you on various public transportation options available to you in your city and helps you navigate from point A to B using different forms of public transportation.

Carbon Offsetting Your Next Flight

According to estimates by the Environmental Protection Agency, aircraft contribute roughly three percent of the United States' total carbon dioxide emissions and 12 percent of the transportation sector emissions. In spite of all of the hubbub about flight delays and shoddy customer service, the Federal Aviation Administration projects air transportation to increase significantly, driving up emissions from domestic aircraft by over 60 percent by 2025. As we noted in chapter 7, however, releasing emissions at high altitudes increases their impact.

States like California have begun asking the EPA to adopt regulations to control greenhouse gas emissions from aircraft. There are currently no such controls for planes and only limited controls on conventional pollutants such as carbon monoxide. So what can individuals do in the meantime?

One approach is to consider making your next flight carbon-neutral. You can calculate the amount of greenhouse gases that your flight generated, and purchase carbon offsets that reduce emissions by that same amount.

Carbon offsets are simply certificates or "shares" that you buy in a project that, for example, plants trees to offset the emissions released by vehicles. Trees have a natural capacity to absorb carbon from the atmosphere. Other carbon-offset programs involve promotion of energy efficient appliance or development of renewable sources of energy.

The website ecobusinesslinks.com offers a comparison of various online carbon offset programs and the types of projects they invest in. Visit www.ecobusinesslinks.com/carbon_offset_wind_credits_carbon_reduction.htm.

Some airlines have implemented carbon offsetting programs. Travelers can view the carbon footprint of their flights and purchase corresponding carbon offsets via contributions to programs like Sustainable Travel International (www.carbonoffsets.org).

Some carbon offset programs have come under fire for allegedly not delivering on the projects contributors are supposedly funding, so it's worth finding out about a given offset scheme to make sure it's legit before you commit.

Recycling

Recycling has been around for decades, and chances are you're already familiar with it in many forms. Newspapers, plastic, aluminum cans—many of us don't have to go further than the curb to recycle these materials. In almost all cases, it takes less energy to produce one of these items from recycled members of its family than from raw materials. For example, a ton of paper made from recycled fiber saves up to 17 trees and uses half the water involved with conventional paper production. For aluminum, the savings are outright huge—aluminum derived from scrap uses 95 percent less energy than mining and processing bauxite ore into the finished product. It's a no-brainer, and today it's big business.

One area of particular interest in recent years has been so-called *e-waste*. Ever wonder what happens to your PC when you replace it with a new one? Well, you might not realize it, but there's enough gold, copper and other metals in consumer electronics devices to make it profitable to ship old computers, cell phones and other gadgets halfway around the world to be picked apart so the metal can be collected and reused.

Energy Efficiency on a Broader Scale   **21**

I N THE PREVIOUS CHAPTER WE FOCUSED on energy efficiency concepts inside the four walls of our house. Now we'll zoom out of the confines of our home and look at energy efficiency from a much broader perspective.

How do we get a community, city, a state, or even the entire nation to adopt these ideas? Is the energy efficiency gospel better spread through a community outreach effort or should government get involved in some way? What about the role of large corporations? Where do they fit in, if at all? And what about the utilities themselves? What programs can they launch that help reduce energy waste among the millions of customers they typically serve?

Conservation and Efficiency on a Larger Scale

A variety of innovative products are being offered in the market today that will allow you to have the same or better quality of life but with a much smaller energy footprint. The private sector understands that energy-sipping devices make good economic sense in addition to helping save the environment. In addition, there are government and utility programs and incentives to stimulate parts of the market in addition to what would have happened naturally.

Many electric utilities are now doing their part to educate us to reduce energy waste as well. They have been running various energy conservation and efficiency programs for the past three decades (at least), ranging from simple contracts with large factories to reduce or cut off their supply in an emergency to sophisticated "demand side management" programs that do essentially the same thing but with more precision.

Government, both at the state and federal level, offer various subsidies and incentives to encourage adoption of energy efficiency. The recent economic stimulus package from the federal government has a lot to offer for the energy industry.

In subsequent sections, we will sample products, initiatives, and innovative programs offered by the public and private sector that will change the way you look at energy, conservation and efficiency. But first, we'll start with an example of a rather quirky way to get the word out on efficiency.

The Energy Smackdown

Most of us probably feel like we should reduce our energy consumption at times, but even if we've taken a few basic steps (e.g., buying an ENERGY STAR fridge, putting in your first few compact fluorescent bulbs), we could probably use some motivation to take a few more. How would you like a camera crew following you around for a few weeks to monitor your every energy move and broadcast it to the world on reality TV? Think that would be, uh, motivating?

That is exactly what three families in a suburb north of Boston agreed to do. *Energy Smackdown* is a reality show on local cable TV in which contestants compete to shrink their carbon footprint in a kind of energy version of *The Biggest Loser*. The show was so popular that a sequel called *Green Streets* is under way in Britain. To find out more about this novel approach, visit www.energysmackdown.com.

Utilities Do Their Part

Utilities and tobacco companies are perhaps the only two industries in the entire universe that actively work to discourage use of the very product they create. (Imagine long distance phone companies encouraging you to cut back your time on the phone, or the restaurant industry encouraging you to eat at home more often.)

Cigarette companies began promoting tobacco-free living after prolonged protests and arm-twisting from the public and the media, not to mention a few attorneys general. Utilities, on the other hand (and some more than others), have been promoting energy conservation and efficiency for many years as a normal course of business. As is often said within the utility industry "A megawatt saved is the equivalent of a megawatt produced."

We'll sample a few programs that they have been running for nearly thirty years now.

Voluntary time-of-use electricity program: This program encourages customers to reduce electricity use between the peak hours, typically between 10 am and 10 pm. Customers who sign up for this program pay an off-peak price that is between 50 to 80 percent less than on a regular electricity pricing plan. Conversely, the rates for usage during the peak time could be much more expensive than the alternative. This is to encourage energy usage during off-peak hours (the carrot) and penalize users for doing the same during on-peak hours (the stick).

Demand-Side Management (DSM): Also known as *demand response*, these programs are designed to literally manage the demand side of the power equation by getting people to alter their energy consumption with an eye toward deferring the construction of new power plants.

The decision to build a new power plant is closely tied to how much energy is consumed during peak time. Sharper peaks call for adding more supply to meet the electricity needs of that peak, typically natural gas power plants that might sit idle most of the time. On the other hand, flatter peaks allow the energy needs to be met by the existing fleet of power plants and with greater predictability for the utility.

Interruptible load program: In this program, willing homeowners allow the electric company to shut off their electric supply during the most critical peak periods when the utility is struggling to meet the needs of all customers in their service territory. In exchange, these "interruptible" customers pay significantly lower rates for electricity than regular customers. This program is sometimes known as the "voluntary demand response program" as it offers consumers incentives to voluntarily reduce their electric loads at system peaks.

Automatic load interruption program (or direct load control): In this program, homeowners agree to install a remote sensor on a specific home appliance, typically a water heater or central air conditioner. When the utility wants to reduce energy consumption across the system, it sends a signal to thousands of devices that will automatically cut the power supply off to the appliances they're connected to for a given length of time. These types of programs allow the utility to instantaneously reduce energy consumption (also called energy demand) across the system.

Energy efficiency programs and incentives: In recent years, utilities have begun offering a variety of innovative energy efficiency programs ranging from a free energy audit of your home to rebates for replacing your old refrigerator or appliance for a new Energy Star compliant one. Similarly, lighting retrofit programs provide assistance with upgrading light fixtures to more energy efficient technologies. Other programs offer to replace existing electric meters with smart meters that allow energy consumption data to be recorded by the utility via Internet connections.

Browse the Internet site of your local utility to better understand the various programs and rebate offers that you can take advantage of.

Energy Efficiency as an Alternative to New Power Generation

Many electric utilities see energy efficiency programs and demand side management programs as an alternative to adding expensive new power plants to the mix. Demand side management programs are typically of two main types.

The first one is where the energy consumption of customers is forcibly reduced or cut-off completely for grid reliability reasons. An example of this would be a brownout program or a rolling blackout program such as the one used in California during the energy crisis of 2000 and 2001. Such actions are seen by the utility as a last resort that is only used to avoid a much larger and more disruptive outage.

The second type of program is one where customers are given economic incentives to reduce their electricity consumption during times when it is cheaper to reduce the demand side of the equation than it is to purchase expensive peak electricity from another utility or to generate additional units of electric supply from a prohibitively expensive power plant.

The Government's Role

One of the most important roles that government can play is that of a catalyst that spurs widespread adoption of energy efficiency "best practices." It does this in many ways, but we'll quickly look at a few.

Establishing standards: The government establishes standards for the energy efficiency of a very wide range of equipment and appliances. Businesses innovate, build, and sell the appliances and devices that conform to these high standards. Household and business consumers purchase the energy efficient appliances and devices. It's that simple.

Of course, the process could get derailed in a variety of ways—lobbyists could water down the regulation, businesses could fight the creation of these new appliances as they could be more expensive (read chapter 8 on CAFE standards for vehicles), and consumers could be slow in adopting more efficient equipment. Just look around your own home, and you'll probably see several incandescent light bulbs still hanging around. If you've changed them all out for compact fluorescents, give yourself a pat on the back.

Providing tax credit and subsidies: Governments at the local, state and federal level can provide incentives to make new energy efficient appliances and devices more affordable. As a matter of fact, they already do that, and the many programs are collected in the Database of State Incentives for Renewables and Energy Efficiency (www.dsireusa.org).

Energy efficient mortgages: These mortgages (also called "EEMs") are part of the federal government's initiative to help us reduce our energy use and minimize our greenhouse gas emissions. An EEM allows the borrower to roll in efficiency upgrades into their mortgage. You can add up to 15 percent of the total mortgage amount this way. You can also access the program via refinancing your home if you make similar efficiency improvements.

Federal tax credits for energy efficiency: The federal government offers a variety of tax credits for energy efficiency. You can find an entire section of the website devoted to these issues at http://www.energystar.gov/index.cfm?c=products.pr_tax_credits.

Federal economic stimulus bill: On February 17, 2009, President Obama signed the American Recovery and Reinvestment Act of 2009, which made some significant changes to energy efficiency tax credits.

The tax credits previously effective through 2009 have been extended to 2010, and those that were for a specific dollar amount have been converted to 30 percent of the cost of the given product/system/upgrade. The maximum tax credit has been raised from $500 to $1,500 for the 2009–2010 period. Up-to-date information on programs and incentives offered by the federal government can be found at http://www.energy.gov/energyefficiency/index.htm and at www.energysavers.gov/.

Daylight Saving Time: Saving Time, Saving Energy

In the past, industry and the government have come together to help our country save energy. Can you guess what that is? (Hint: It has to do with the phrase "spring forward, fall back.")

Yep, daylight saving time (DST). Most of us delight in getting that extra hour of sleep in fall, and detest losing it again in spring. But did you know that daylight saving was implemented primarily as an energy saving measure?

New start and stop dates for DST were established by the Energy Policy Act of 2005, but the origins of daylight saving go back nearly a century.[1] Time zones were first created in the United States by the railroad companies to standardize their arrival and departure schedules. Until then, clocks were set by the stars (no joke). In 1918, the U.S. Congress adopted the rail zones as federal law.

Modern DST was first proposed in 1907 by the English builder William Willett who was reportedly passing by London homes where the shades were down, even though the sun was up. Willett proceeded to write a document titled "The Waste of Daylight."

In the U.S, DST was enshrined in law by the Uniform Time Act of 1966. It doesn't require that anyone observe daylight saving time, only that if you are going to observe it you have to do it in a uniform manner.

The premise for following this practice is tied to saving electricity usage, in particular, savings from reduced usage of residential lighting because of availability of natural sunlight. This is how it works.

As noted earlier, human activity works in cycles and so does the energy consumption in the average household. On an average, over 25 percent of our monthly energy consumption comes from operating our light bulbs, and small appliances such as TVs, DVD players, personal computers and music stereos, and much of that consumption occurs in the evening. Delaying the time of sunset and sunrise reduces the use of artificial light in the evening and increases it in the morning. Advocates of daylight saving maintain that the savings in the evening are far greater than the energy increase in the morning.

Following the 1973 Arab oil embargo, the U.S. went on extended daylight saving time for two years in hopes of saving energy. The experiment worked until 1975 when it was discontinued because of opposition, mostly from farming states. To this day, this practice is controversial in that the actual energy savings are contested by some experts. However, over 70 countries around the world continue to observe daylight saving time in one form or another.

Corporations Go Green: Over-Hyped or a True Phenomenon?

"Greenwashing" is a term applied to corporations' attempts to appear environmentally conscious without making substantive changes in the way they do business. With all the hype that the whole green movement has received, it would be impossible for all of it to be founded on real change. It seems like everyone is getting into the act. There are even "green shoots" for TV shows. MTV's ever-popular show, *The Real World*, now comes to us from an "eco-friendly crib." Yeah, no kidding. The house is tricked out with efficient appliances, loads of recycled materials, and even a hot tub heated by the sun.

Having said that, there are many corporations that have made impressive strides to improve the environmental friendliness of their products. General Electric, ever the master of marketing, launched a new ad campaign around the trademarked word "ecomagination" to highlight the company's green products. Wal-Mart is putting solar panels on the roofs of some of its big-box stores. The list of corporate environmental good deeds is getting longer by the minute.[2]

Beyond the ad-speak, though, corporations are in a position to make a huge impact in energy conservation and efficiency.[3] Some firms have signed power supply contracts to get all their electricity from renewable resources. One report put the total at 5 billion kilowatt-hours. Other companies are taking a microscope to their operations to find ways to improve their energy use and their productivity at the same time.

One example comes from Bob's "day job" at ABB, which is, among other things, the world's largest supplier of electric motors. The funny thing about most motors in industrial plants, though, is that they have only one speed: maximum. They're either running flat-out or they're off. Adding a device called a drive to control the motor's speed can save anywhere from 15 to 70 percent of the energy used to run the motor, but currently only a small fraction of the world's electric motors are outfitted with variable speed drives.

Utilities, too, are improving their efficiency behind the scenes, in addition to the many incentives and other programs they offer for their customers. The industry is now in the process of implementing higher efficiency standards for distribution transformers (of which there are millions on the U.S. grid). The improvement (reduction) in energy losses is fairly small on a per-unit basis, but when you add up all those transformers, the energy savings are significant.

Clearly, the potential for reducing our energy consumption, not to mention our energy costs, is enormous. And if there's one thing corporations are good at it's finding ways to reduce costs. That's the beauty of efficiency—it comes with a ready-made business case.

So, Where Do We Go from Here?

Our goal with this part of the book was not to point fingers or assign blame for how inefficient our current energy systems are, but rather to point to the bright side. There are many, many ways that individuals, businesses, and institutions can drastically improve their energy consumption using proven technologies that are already widely available. In most cases, if not all, saving energy also means saving money, at least over the long term.

Admittedly, energy efficiency is not a particularly sexy topic, and there is no one blockbuster technology or practice that can be applied across all kinds of energy usage. But as we noted at the beginning of this part, energy efficiency is by far the most expedient way to reduce our energy consumption and in the process reap all the economic and environmental benefits that come along with that. It comes down to a large number of small steps that together add up to significant change. In the preceding chapters, we've touched on just a few ways to improve energy efficiency, but there are many more. We could write a whole book on the subject, but don't wait for us—check out your local bookstore or look online for one of the numerous books that give practical advice on how to improve the efficiency of your home or business.

A NEW ENERGY ECONOMY V

T HE INTERNET STARTED OUT AS A simple idea: link multiple computers together so they could share information between them. It existed for years as something of a technological backwater, known only to certain military and scientific research communities, but with the arrival of the World Wide Web in 1993, everything changed.

The Internet soon became the enabling technology behind a revolution that swept across numerous other industries from finance to retailing and of course communications. Today, it's hard to identify a single industry that hasn't been affected by the Internet. Indeed, there probably aren't any.

The Internet revolution also made a lot of people rich, and we're not just talking about all those goofy dot-com enterprises that made a couple of millionaires before flaming out. Savvy investors who saw the potential in online retail (Amazon), search engines (Google), or online auctions (eBay) made bundles.

Today, the smart money is starting to flow into a new revolution: clean technology, or "cleantech" as it's often called. This is the intersection of commerce and climate change, energy, and sustainability, though this revolution has already had a major shakeout thanks to the global recession that arrived in 2008. Still, the potential is huge. Weaning our global economy off of carbon will require a massive shift, primarily in the energy industry, and any product or technology that can deliver real results at a reasonable price is likely to get the attention of investors.

In this part, we'll take a quick look at the future of energy in the context of investing. We're not giving advice on where to put your nest egg, just taking a peek at where the next Google might come from.

Cleantech Investments 22

IN JANUARY 1848, JOHN SUTTER was building a sawmill on the American River in Coloma, California, near Sacramento.[1] James Marshall was running the work crew one day when he noticed a few tiny specks glittering in the water. It wasn't long before word got around that there was gold in the American River, and in May a local shopkeeper named Sam Brannan produced a bottle full of gold dust he'd collected from his customers as payment. The Gold Rush was on.

The Internet boom and cleantech have both been likened to the Gold Rush, though it's worth pointing out that some of the people who made out the best in the 1800s never worked a pan or wielded a pickax. Nope, guys like Brannan figured out pretty quickly that finding gold might be hit and miss but supplying all those would-be miners was a sure thing. Today, several of San Francisco's streets are named for these entrepreneurs, Brannan among them.[2]

The cleantech gold rush, to be fair, is taking a somewhat different course than even the dot-com boom. As we write this, we're in the middle of a recession that has put the brakes on investments of all kinds, including cleantech, and a number of solar companies and others have already gone bankrupt. Still, there is no doubt that our efforts to trade our dependence on hydrocarbons for renewable energy and greater efficiency present a tantalizing opportunity.

Cleantech Investing: Wall Street and Silicon Valley's New Love Affair

Initially, most of the attention from investors and entrepreneurs in this "space" (to use a dubiously chic term) was focused on the obvious: hybrid cars, solar and wind power, biofuels, and an assortment of efficiency-boosting technologies. More recently, the cleantech umbrella has opened wider to include things like smart grid technologies, water purification, and even services like carbon offset programs.

However, most cleantech businesses are highly capital-intensive, so you're not likely to find very many of the proverbial garage operations out there. If you have a good idea for a new kind of thin-film solar panel, for example, you'll have to come up with many millions of dollars before you can even start making prototypes for testing. In fact, according to Lux Research, the average cleantech startup will require 2.5 times as much money as a comparable information technology business to get off the ground.[3]

That isn't stopping the movers and shakers from, well, moving and shaking in cleantech. One contributor to the website altenergystocks.com compared the sector to the most influential technology/industrial movements in history, right up there with steam power and the automobile.[4]

So What's Driving the Investment?

Remember *Wall Street*? You know, the movie with Charlie Sheen as the naive stockbroker and Michael Douglas as the "greed-is-good" corporate raider? You might recall the advice Sheen's character, Bud Fox, gets from his dad near the climax of the picture. Carl Fox (played by Sheen's real-life dad Martin) implores the young buck to "create, instead of living off the buying and selling of others."

Well, the cleantech world has enough room to accommodate the creators as well as the buyers and sellers, because this is one industry that has a bit of everything in it. So what's making it go?

First, there is the perfect storm of climate change, national security, and long-term viability of the U.S. economy—all of them point to a need for a cleaner energy supply, and smarter ways of using it. Then there is simple supply and demand economics. Oil is (historically) getting more expensive as it becomes scarcer and demand for it continues to increase. That alone will provide an incentive for the development of alternative fuels.

We're also in need of some maintenance on our existing energy infrastructure, the power grid in particular. So, it seems only reasonable to improve its sustainability while we're upgrading its capacity and reliability.

Cleantech is also an industry of tangible things (for the most part), and as such it's easier to understand than the often arcane business models of information technology companies. There's also the obvious allure of rebuilding America's manufacturing capability. Wind turbines, solar panels, and carbon capture plants don't install themselves, and the jobs they create aren't easily outsourced. (They might be in-sourced, meaning imported, but the point is that the work gets done here—deciding who does it is another matter.)

Finally, there's the government, which exercises great influence in two ways. First, it sets energy policy, which lately has meant an emphasis on encouraging clean energy, but it also makes investments of its own. The American Recovery and Reinvestment Act of 2009 (ARRA), for example, has a multitude of provisions for grants and other funding mechanisms to spur the development and implementation of cleantech. All that public money sloshing around is bound to attract some attention.

We should also note the appeal cleantech has in terms of being a do-good investment. Gordon Gekko (*Wall Street* again) would probably scoff at the notion, but given the rise in popularity of investment vehicles like socially responsible mutual funds, it's clear that there is a market for investments that offer benefits beyond a return in dollars.

Cleantech[5] investment is booming. According to data from Thomson Reuters and the Cleantech Group, investments in the sector have roughly doubled every year from 2005 to 2008, despite the onset of the "Great Recession."[6] So, how big can this market get? Well, we are dealing with "the world's most vital commodity," not to mention the sustainability of our modern lifestyle. How much is that worth?

To put some numbers on it, though, a company called New Energy Finance has posted a figure of $155 billion as the amount of money that has already flowed into cleantech ventures worldwide.[7] Oh, and that was just for 2008. That sounds like a lot, but the research firm also says that in order to meet our objectives for stabilizing atmospheric CO_2 levels, we'll need to pump more than half a *trillion* dollars per year between now and 2030.[8] The venerable International Energy Agency puts the figure even higher at $800 billion per year.[9, 10]

Here in the U.S., cleantech investing by venture capital and private equity firms totaled $13.5 billion in 2008.[11, 12] Again, this is in the middle of the worst recession in 75 years. Industry reports also paint quite a picture looking forward. One analysis concludes that if the market for biofuel-driven cars doubles to 6 percent of the total market over the next four years, that alone would create a $28 billion industry almost overnight.[13, 14, 15]

Uncle Sam Invests in a Cleantech Future

Given the scope of the investment needed to realize even an energy supply green enough to stabilize our CO_2 levels at the 450 ppm that scientists have pegged as a maximum,[16] you might ask where those hundreds of billions are going to come from. The private sector has vast resources, but some of them will stay on the sidelines until there's a clear path to profit. Getting from here to there will take more than the market. It's going to take—cue the music of your choice—the government.

From the environmentalist's perspective, this fact is a matter of practicality. We simply don't have the time to wait for clean energy technologies to grow on the open plains of the free market, so for the sake of speed alone we need—wait for it—a greenhouse, where they can take root.

From a more nuts-and-bolts economic viewpoint, government intervention in the form of tax incentives, loan guarantees, and direct subsidies will allow cleantech businesses to ramp up. The idea is to get to a point where economies of scale and simple experience bring costs down to a point where governmental support is no longer needed. Wind power, which we discussed back in chapter 1, is probably the best example of this in the U.S. The cost of wind power has dropped dramatically over the past twenty years, despite several interruptions in support. And while it's still a long way from being competitive with coal or nuclear, wind is now competitive with gas-fired generation when natural gas prices are high.

So, entrepreneurs, investors, and Uncle Sam are going to have to play nice together if the grand predictions of a new energy economy are to be fulfilled.

Fun Fact: Did You Know?

About $50 to $65 billion of the $731 billion American Recovery and Reinvestment Act is tagged for energy efficiency, renewables, and the like.[17, 18] That's a little shy of 9 percent of the total. But the U.S. isn't the only country funneling serious money into cleantech as part of a larger economic recovery effort. China is slated to spend a thumping $221 billion on clean energy.[19]

The task before the energy industry is so huge, it's hard to grasp. Here is an industry, already the largest the world has ever known, and now it needs to grow even larger while fundamentally changing its very nature. No small task, and if you're skeptical about whether we're even up to it, we can't blame you, but in challenge lies opportunity, to be sure.

Energy, Economy, Jobs, and Education

23

VAN JONES IS A SOCIAL ACTIVIST turned environmentalist turned policy advisor. Before he found himself in a Washington dust-up that forced his resignation, he served as the Special Advisor for Green Jobs, Enterprise, and Innovation at the White House Council on Environmental Quality. That's a big title, but it's a big job.

Jones is also the author of a best-selling book called *The Green Collar Economy: How One Solution Can Solve Our Two Biggest Problems*, which blends his interests in social justice and environmentalism. His main thesis: Making our society more environmentally sustainable is a great way to boost the economy and lift people out of poverty. We'd say he's on to something, but he's not the only one. "Green jobs" are all over the news these days, and with good reason. As the subtitle to Jones's book indicates, the idea of job creation in the field of clean energy has a broad appeal. It's even more appealing now, as the world swoons from a global recession that has unemployment in the U.S. hovering around 10 percent for the first time in recent memory.[1]

In this chapter, we'll take a look at what "green jobs" are all about, and how energy and economics come together in the new energy economy.

Green Jobs and a Clean Energy Economy

Actor Jim Carrey has had a couple of "green" jobs. He donned the color in 1994 for *The Mask*, and again in 2000 for *How the Grinch Stole Christmas*. For the latter film, he is reported to have spent hours in the makeup chair every day for 92 days of shooting, and for that the film took home an Oscar for best makeup. But for our purposes, "green jobs" refers to what you do as opposed to what color your body is when you do it.

So, what qualifies as a green job? The term has expanded in recent years, actually. The idea was first applied to manufacturing and other jobs directly associated with clean energy, like building wind turbines or manufacturing solar panels, but now it encompasses a broad range of occupations. In addition to the manufacturing jobs, it includes everything from installing insulation to writing software to manage a building's energy use. Green jobs don't necessarily have to be directly associated with energy, but many of them are for the simple reason that energy is so integral to making or economy more environmentally sustainable—more "green."

The manufacturing segment of green jobs is especially important for the U.S. because those kinds of jobs have been on the decline for decades. Media reports tell us we're becoming a service society that doesn't actually make anything anymore, and there's a good deal of truth in that. Rust-belt states like Ohio and Michigan, once home to tens of thousands of manufacturing jobs, have seen their local economies wither, along with the cities that used to be supported by them. Detroit is the poster child for this. The city now resembles a doughnut, a ring of suburban life surrounding a hollowed-out city center.

Green jobs in manufacturing are just the beginning, though. What has captured the attention of policy makers and other interested parties is how very local a lot of these jobs are (i.e., they cannot be shipped overseas). For example, if improving energy efficiency in homes and commercial buildings is the quickest way to make a dent in our energy consumption, well, the work that needs to be done is right there inside of every house, school, office park, and factory across the country. Weatherization of homes could employ hundreds of people even in a small city, given the right incentives.

We should also note that green jobs cover a wide range of skill sets and education levels. Take solar power. We need scientists to develop new materials and more efficient PV cells. We also need engineers to design the panels, factory managers to run the plants, and an army of installers to put the finished product on rooftops.

It's worth noting, as we did earlier, that there's nothing to stop a solar installer, for example, from hiring a bunch of workers from outside the U.S. (assuming they've passed muster with Immigration and Customs Enforcement). The idea is simply that the work itself is *here*.

Growth in green jobs will benefit greatly from legislation that puts value on energy efficiency and a price on carbon, so there is a policy angle here, and the government has a big role to play. One example might be a renewable energy standard (RES), which as we discuss in Volume 1, is a mandate for utilities to generate a certain percentage of their power from renewable sources. Many states already have an RES, and one by-product of those laws is new demand for new wind farms, solar installations, geothermal plants, and so on.

So how many green jobs can we make? One study released by the Blue Green Alliance and the Renewable Energy Policy Project looked at the effect a national RES would have on employment. The study found that such an effort aimed at bringing 185,000 MW of renewable energy online over ten years would generate 850,000 jobs at a cost of $160 billion.[2] Another study found that $100 billion in clean energy spending could produce 2 million jobs, and in only two years.[3]

Obviously, this kind of prognostication is not an exact science, and for every study put forward by organizations with the word "green" in their name there is another less optimistic one published by another interest group of one kind or another. But there is certainly an elegant logic to the notion of boosting the economy by putting people to work on things that will reduce pollution, cut CO_2 emissions, reduce our dependence on foreign sources of energy, and (eventually) lower energy costs. That may be what moves the needle.

Energy Education

So OK, all this green jobs stuff sounds pretty good, but are we really prepared to lead the world in clean energy? That depends. The U.S. isn't the only nation with great universities, a vibrant business community and government incentives for clean energy. If you think of cleantech as a market, we are but one competitor, and achieving a leadership position is going to require a highly skilled workforce. All those different kinds of green jobs we listed a moment ago need qualified people to fill them. We need education, specifically energy education, to ensure we have the resources (human and otherwise) to meet the challenge.

Now, given the fact that our stated purpose with this book is to raise the level of energy literacy among regular people, we obviously have some skin in this game. So, consider this full disclosure: we want everyone to understand energy to the extent that it affects their lives. Not everyone has to be a nuclear engineer or a fuel cell researcher, but we do think it's important for everyone to have a basic idea of where their energy comes from, what it costs, and why it is so important.

Reading this book will help—a little—but the education we're talking about (OK, advocating) here is more comprehensive. Third graders need to know where electricity comes from so that when they grow up they'll have some idea of what they're paying for every month on their utility bill. The more we know, the better able we'll be to make informed decisions as consumers, which ultimately is probably the most effective way to make the shift to a sustainable energy future.

The Coming Energy Revolution   **24**

Eckhart Tolle is a German-Canadian spiritual teacher, motivational speaker, and author who's achieved that pinnacle of every writer's dreams—appearing on the Oprah Winfrey show. (We're not holding our breath, but you never know—this energy thing could catch on.)

He opens his book *A New Earth: Awakening to Your Life's Purpose* with an account of the very first flower to open on planet Earth and goes on to describe a groundswell of activities leading to an eruption of flowers across the globe. It's a useful analogy here because the changes that are coming our way with respect to energy are like that first flower, a revolution.

The actual process, though, will be more like evolution—gradual changes, each building on the ones before and all of them directing us toward a better shot at survival.

If you've read through all the other chapters in this book, well, there isn't much more for us to tell you. Your introduction to energy is just about complete. In these final pages, we'll take a look at how the coming energy revolution/evolution will show up in a variety of different aspects of our everyday lives.

The Coming Energy Revolution

Manufacturing. We've already talked about how a boom in cleantech could spur a revitalization of America's manufacturing sector, but in truth it's already starting to happen. Wind turbine makers from Europe have had their sights set on the massive U.S. market for years, but have stayed away largely because of the lack of certainty in key policies, namely the production tax credit (PTC) for wind. Now that the PTC has been renewed for another few years and it looks like there's a good chance it will be extended even further, turbine makers are setting up shop in Pennsylvania, Colorado, and elsewhere.

Watch too, for more manufacturers of other products to make improvements in the energy efficiency of their plants, and even to generate their own power through cogeneration or even solar panels on the roof.

Real estate. Buildings are the biggest source of CO_2 emissions, and most of the ones standing today will still be here twenty to fifty years from now. That makes for a huge market in efficiency retrofits. Meanwhile, the U.S. Green Building Council's Leadership in Energy and Environmental Design (LEED) standards are being adopted by an increasing number of builders. LEED-certified buildings made up about 5 percent of new commercial space in the U.S. in 2008,[1] but that figure is expected to rise significantly in the coming years. Residential construction is also changing. "Green homes" are now fashionable and according to a recent report will make up 10 percent of new home construction in 2010.[2, 3]

Corporate social responsibility. CSR doesn't just mean "customer service representative" anymore, and major corporations are taking their "social footprint" seriously. Not all that long ago, an alliance between the likes of Caterpillar, BP, and the Natural Resources Defense Council would have sounded like a punch line. In fact, they are all members of the U.S. Climate Action Partnership (USCAP),[4] one of the most prominent climate change organizations. Expect more bridging of old divides as the idea of sustainability is adopted more widely and businesses see benefits to going green beyond simple enhancement of their public image.

One example of this is the Climate Savers Computing Initiative, a non-profit group committed to development and implementation of energy-saving technologies in electronic devices.[5] The group has set a goal of cutting CO_2 emissions from electronics by 54 million tons per year by 2010.

Agriculture. We talked a bit about the food vs. fuel debate in chapter 10, and the questions surrounding corn-based ethanol in particular are not likely to go away anytime soon. But aside from questions about cropland being used for biofuels, the farming industry itself is ripe for an energy makeover. Food production on a mass scale is highly energy-intensive, but farming enjoys a certain amount of restraint in government regulation over pollution associated with farm operations. Air quality in some farming communities, particularly in California's Central Valley, is among the worst anywhere in the U.S. in part because farm equipment isn't subject to the same emissions controls as cars and trucks. New regulations could change that, but it will come at a price, one we'll all pay at the grocery store.

Biotech. The biotech industry will have a major role to play in the development of biofuels and the plants that are used to make them, but it might also be instrumental in carbon capture. Demonstration projects are already up and running to see if carbon-eating algae can be used to keep the CO_2 from coal-fired power plants from ever reaching the atmosphere. If it works and can be scaled up to service the likes of a 1,200 MW coal plant, it could prove to be a critical step in weaning us off coal or at least mitigating its environmental impact.

Retail. Say what you will about Wal-Mart, the undisputed king of retail is in a unique position to drive change among its thousands of suppliers through initiatives like buying organic products and insisting on more environmentally friendly packaging. Other major retailers are finding a hidden resource in (or rather on) their big box stores—the roof. Many big box retail stores are located in wide-open areas with no trees or tall buildings shading their roofs and several grocery stores have the same situation. This makes them ideal sites for solar power. Companies as diverse as Home Depot, Whole Foods, REI, and Wal-Mart are experimenting with rooftop solar power to meet at least a portion of their electricity requirements.

Media and lifestyle. We grouped these two things together because it seems they are increasingly linked. Even way back at the dawn of the Internet Age, a new magazine was launched that had as its tagline the somewhat (at the time) incongruous phrase "technology as lifestyle." The publication in question was *Wired* magazine, and since then media has become a never-ending feedback loop of personal style and culture focused on the individual. And with environmental issues taking up so much of the scene these days, is it any wonder that MTV's original reality show *The Real World* now puts the participants up in green homes with energy-efficient appliances and on-site composting? California Governor Arnold Schwarzenegger even appeared on an episode of *Pimp My Ride* to talk up biodiesel vehicles. Now you know you're not in Kansas anymore.

Government. We've talked a lot in this book about the role of government in directing our energy future, and we're finishing this section with it because it is so vitally important. Energy policy at the federal, state, and local level will determine perhaps more than anything else what our energy future will look like. But aside from tax credits, emissions caps, and the like, the government is also an energy consumer—the largest in the world, in fact—and as such it wields tremendous influence in the marketplace. If Uncle Sam decides he's only going to buy compact fluorescent light bulbs or Energy Star appliances from here on out, you can bet someone will be there to sell them.

But government, at least in this country, is not a spectator sport. The more engaged we all are in the process, the better our energy policy will be and the less time we'll spend wrangling over ideology and minutiae. If you want to find out more about what's going on in your area, check out the list below of some noteworthy initiatives at the state and local level.

STATE LEVEL INITIATIVES

- Green California
  www.green.ca.gov
- Illinois Green Government Council
  www.standingupforillinois.org/green/
- New York State Green Building Initiative
  www.dec.ny.gov
- North Carolina Project Green
  ww.ncprojectgreen.com
- Pennsylvania Green Government Council
  www.gggc.state.pa.us

COUNTY

- Santa Barbara County Green Award Program
  www.lessismore.org

CITY

- City of Los Angeles
  www.ladwp.com/ladwp/areaHomeIndex.jsp?contentId=LADWP
  _GREENLA_SCID
- City of Palo Alto
  www.cityofpaloalto.org/environment/default.asp
- City of Santa Monica
  santa-monica.org/EPD/EnvironmentalPrograms
- City and County of San Francisco
  www.sfenvironment.com
- City of Portland
  www.sustainableportland.org/
- City of Seattle
  www.seattle.gov/environment/
- City of Oakland
  www.sustainableoakland.com

Epilogue

CONGRATULATIONS! YOU'VE MADE IT to the end, or the end of the beginning anyway. We hope you've enjoyed the trip, and that you've come away with a better understanding of energy issues and why this subject is so important. We also hope you had some fun along the way. We set out to make energy understandable with a healthy dose of humor because, well, you just can't be that serious all the time, even about a topic as serious as this. Energy is important, but it needn't be boring.

This doesn't have to be the end of the line, though. Maybe you've just pocketed a few conversation tidbits for your next cocktail party—and that's OK—but there are many ways you can put your energy savvy to work. Just being an informed consumer is one. And if you're moved to take further action (e.g., drive a hybrid, put solar on the roof, quit your job and go work for an energy non-profit), that's great too.

For our part, we're just getting started. We're on a mission to increase energy literacy and this book is only the first step. If you have something to say about that—what to include in our next book, for example—we'd love to hear from you. Look us up online at www.energyexplained.com.

Meantime, keep on with whatever got you to pick up this book in the first place. Woody Allen famously said, "eighty percent of success is just showing up," and we're glad you showed up for our little visit to the world of energy. We hope to see you again sometime soon.

Notes
Volume 2: Alternative Energy

Chapter 1: Wind Energy

1. Wind Turbines and Windfarms Database, "Generalities about Wind Energy," The Wind Power, www.thewindpower.net/topic-51-generalities-about-wind-energy-1.php.

2. California Energy Commission, "Wind Energy," Energy Quest, www.energyquest.ca.gov/story/chapter16.html.

3. Energy Information Administration, "History of Wind Power," http://tonto.eia.doe.gov/kids/energy.cfm?page=wind_home-basics.

4. U.S. Department of Energy, "History of Wind Energy," September 12, 2005, www1.eere.energy.gov/windandhydro/wind_history.html.

5. "Exploring Wind Energy—Student Guide," The Need Project, 2009–2010, www.need.org/needpdf/Exploring%20Wind%20Student%20Guide.pdf.

6. American Wind Energy Association, "Basic Principles of Wind Resource Evaluation," www.awea.org/faq/basicwr.html.

7. U.S. Department of Energy, "2% Wind Energy by 2030," July 2008, www1.eere.energy.gov/windandhydro/pdfs/41869.pdf.

8. R. Villafafila, A. Sumper, A. Suwannarat, B. Bak-Jensen, R. Ramirez, O. Gomis, and A. Sudria, "On Wind Power Integration into Electrical Power System: Spain vs. Denmark," Polytechnical University of Catalonia and Aalborg University, www.icrepq.com/icrepq07/374-villafafila.pdf.

9. The installation costs are now creeping up toward $1,500—$1,750 per installed kilowatt.

10. Daniel Yergin, "A Great Bubbling," *Newsweek*, December 4, 2006, www.newsweek.com/id/44178/page/1.

11. Michael Goggin, "Wind, Backup Power and Emissions," *EnergyPulse Magazine*, June 23, 2009, www.energypulse.net/centers/article/article_display.cfm?a_id=2078.

12. World Wind Energy Association, "Wind Energy Market Worldwide Continues Strong Growth," June 23, 2009, www.wwindea.org/home/index .php?option=com_content&task=view&id=245&Itemid=40.

13. Neal Babcock, "Catch the Wind and Crank Up Your Turbine," Engineer and Technician, 2009, www.engineer-and-technician.com/catch-the-wind-and -crank-up-your-turbine/.

14. Walt Musial, Sandy Butterfield, and Andrew Boone, "Feasibility of Floating Platform Systems for Wind Turbines," National Renewable Energy Laboratory, January 5–8, 2004, www.nrel.gov/docs/fy04osti/34874.pdf.

Chapter 2: Solar Energy

1. George Harrison, "Here Comes the Sun," Song Facts, www.songfacts .com/detail.php?id=197.

2. C. Johnson, "How the Sun Works in Making Light and Heat," August 2006, http://mb-soft.com/public2/sunworks.html.

3. Geoffrey Heal, "The Economics of Renewable Energy," *National Bureau of Economic Research Working Paper No. 15081*, June 2009, www.nber.org/papers/ w15081.pdf.

4. Jet Propulsion Laboratory, "The Solar System," California Institute of Technology, www2.jpl.nasa.gov/basics/bsf1-1.php.

5. Rainbow Power Company Ltd, "Solar Panel FAQ," www.rpc.com.au/ products/panels/pvmodules/pvmodules.html.

6. William Pentland, "Solar Energy: A Synopsis of Everything under the Sun," July 30, 2009, http://knol.google.com/k/william-pentland/solar-energy/ 1g0rrsoesmjko/2.

7. National Resource Council Committee on Materials Research for Defense, *Materials Research to Meet 21st Century Defense Needs* (Washington, DC: National Academies Press, 2003), 55–94.

8. Peter Lorenz, Dickon Pinner, and Thomas Seitz, "The Economics of Solar Power," *The McKinsey Quarterly*, June 2008, www.mckinsey.com/clientservice/ ccsi/pdf/economics_of_solar.pdf.

9. National Renewable Energy Laboratory, "Energy Technology Cost and Performance Data," May 2009, www.nrel.gov/analysis/costs.html.

10. Lamar Alexander (R-Tennessee), "The Perils of Energy Sprawl," Floor Remarks in U.S. Senate, October 5, 2009, http://alexander.senate.gov/public/ index.cfm?FuseAction=Speeches.Detail&Speech_id=0a6f9273-5dbc-4c37-99b6 -a9940780c51d&Month=10&Year=2009.

11. National Renewable Energy Laboratory, "Energy Technology Cost and Performance Data," May 2009, www.nrel.gov/analysis/costs.html.

12. U.S. Energy Information Administration, "Electric Power Monthly," December 2009, www.eia.doe.gov/cneaf/electricity/epm/epm_sum.html.

13. Michael C. Baechler, Theresa Gilbride, Kathi Ruiz, Heidi Steward, and Pat M. Love, "Building America Best Practices Series: Solar Thermal and Photovoltaic Systems," U.S. Department of Energy, June 4, 2007, http://apps1.eere .energy.gov/buildings/publications/pdfs/building_america/41085.pdf.

14. U.S. Energy Information Administration, 2007, "Federal Financial Interventions and Subsidies in Energy Markets," www.eia.doe.gov/oiaf/servicerpt/ subsidy2/pdf/execsum.pdf.

15. Michael Schirber, "Whatever Happened to Solar Power?" *Live Science*, December 10, 2007, www.livescience.com/environment/071210-solar-power.html.

16. "Solar Cell Sets World Efficiency Record at 40.8 Percent," *Science Daily*, September 30, 2008, www.sciencedaily.com/releases/2008/09/080929220900 .htm.

Chapter 3: Tidal Energy

1. Gordon Feller, "Wind, Waves, Tides—Ocean Energy," Eco World, www .ecoworld.com/water-supply/winds-waves-tides-ocean-energy.html.

2. U.S. Department of Energy, "Ocean Tidal Power," December 30, 2008, www.energysavers.gov/renewable_energy/ocean/index.cfm/mytopic=50008.

3. Claire Soares, "Tidal Power: The Next Wave of Electricity," *Pollution Engineering*, July 1, 2002, www.pollutionengineering.com/Articles/Feature_Article/ 51b96a005dd68010VgnVCM100000f932a8c0.

4. World Ocean Observatory, "Tidal Energy: Tidal Energy Physics and Resource," www.thew2o.net/events/oceanenergy/images/tidal_energy.pdf.

Chapter 4: Biomass Energy

1. U.S. Environmental Protection Agency, "10 Fast Facts on Recycling," August 8, 2008, www.epa.gov/reg3wcmd/solidwasterecyclingfacts.htm.

2. Idaho Public Television, "Facts about Garbage," Dialogue for Kids, April 19, 2005, http://idahoptv.org/dialogue4kids/season6/garbage/facts.cfm.

3. "Facts and Figures," Dairy Farming Today, www.dairyfarmingtoday.org/ DairyFarmingToday/Learn-More/Facts-And-Figures/.

Chapter 5: Geothermal Energy

1. Encyclopedia.com, "Geothermal Energy," www.encyclopedia.com/ doc/1G2-3451200011.html.

2. Christopher Mims, "One Hot Island: Iceland's Renewable Geothermal Power," *Scientific American*, October 20, 2008, www.scientificamerican.com/article.cfm?id =iceland-geothermal-power.

3. C. J. Bromley and M. A. Mongillo, "Geothermal Energy from Fractured Reservoirs: Dealing with Induced Seismicity," International Energy Agency, *Open Energy Technology Bulletin*, February 2008, www.iea.org/impagr/cip/pdf/Issue48Geothermal.pdf.

4. International Energy Agency, "World Energy Outlook 2009," www.worldenergyoutlook.org/docs/weo2009/WEO2009_es_english.pdf.

5. Kara Slack and Karl Gawell, "Geothermal Energy Association Weekly Update," Geothermal Energy Association, June 30, 2008, www.geo-energy.org/publications/updates/2008/GEA_Weekly_Update_June_30_2008.pdf.

6. Chris Greenwood, Eric Usher, Virginia Sonntag-O'Brien, Alice Hohler, Alice Tyne, Camila Ramos, Fatma Ben Fadhl, Jun Ying, Maggie Kuang, Rohan Boyle, and Sara Lynn Pesek, "Global Trends in Sustainable Energy Investment 2009," United Nations Environment Programme, www.unep.org/pdf/Global_trends_report_2009.pdf.

7. A. Kagel, "A Guide to Geothermal Energy and the Environment," Geothermal Energy Association, 2007, www.geo-energy.org/reports/Environmental%20Guide.pdf.

8. Jefferson W. Tester, Brian J. Anderson, Anthony S. Batchelor, David D. Blackwell, Ronald DiPippo, Elisabeth M. Drake, John Garnish, Bill Livesay, Michal C. Moore, Kenneth Nichols, Susan Petty, M. Nafi Toksoz, and Ralph W. Veatch Jr., "The Future of Geothermal Energy," Massachusetts Institute of Technology, 2006, http://geothermal.inel.gov/publications/future_of_geothermal_energy.pdf.

Chapter 6: Bottle That Electron!
(Energy Storage)

1. Richard Baxter, *Energy Storage: A Nontechnical Guide* (Tulsa, Oklahoma: PennWell Books, 2006), 55–160.

Chapter 7: Land, Sea, and Air—
Energy in Transportation

1. Brian Hagerty, "What Is a Knot? What Is a Nautical Mile?" Online Conversion, 2008, www.onlineconversion.com/faq_07.htm.

2. Jonathan Hiskes, "Tips for Flying to the Copenhagen Climate Conference," Grist.org, May 18, 2009, www.grist.org/article/2009-05-15-tips-for-flying-to-copenhagen.

3. Steven Cherry and Erico Guizzo, "Write and Wrong," *IEEE Spectrum*, November 2004, http://spectrum.ieee.org/energy/nuclear/write-and-wrong.

4. Jeff Hays, "Trains in China," Facts and Details, http://factsanddetails.com/china.php?itemid=315&catid=13&subcatid=86.

5. Stacy C. Davis, Susan W. Diegel, and Robert G. Boundy, "Transportation Energy Data Book—Edition 28," U.S. Department of Energy—Energy Efficiency and Renewable Energy, 2009, www-cta.ornl.gov/data/tedb28/Edition28 _Full_Doc.pdf.

6. Ken Blackwell, "Deregulation Works," National Taxpayers Union, June 12, 2008, www.ntu.org/main/press_commentaries.php?PressID=1024&org _name=NTU.

7. "Fuel 101: From Well to Wing," Air Transport Association of America, Inc., June 15, 2009, www.airlines.org/economics/energy/fuel+101.htm.

8. Chuck Squatriglia, "GM Claims Chevrolet Volt Will Get 230 MPG," *Wired*, August 11, 2009, www.wired.com/autopia/2009/08/chevrolet-volt-230 -mpg/.

9. James Strickland, "Energy Efficiency of Different Modes of Transportation," Strickland's Website, November 25, 2009, www.strickland.ca/efficiency .html.

10. Richard Cobbs and Alex Wolf, "Jet Fuel Hedging Strategies: Options Available for Airlines and a Survey of Industry Practices," Kellogg School of Management, Spring 2004, www.kellogg.northwestern.edu/research/fimrc/papers/ jet_fuel.pdf.

11. Lisa LaMotta, "Southwest Soars Despite Fuel Costs," January 22, 2009, *Forbes*, www.forbes.com/2009/01/22/southwest-airlines-fuel-markets-equity -cx_lal_0122markets12.html.

Chapter 8: Fuels from Fossils (Revisited)

1. Liam Matthew Brockey, *Journey to the East: The Jesuit Mission to China 1579–1724* (Cambridge: Harvard University Press, 2007), 125–164.

2. Robyn Conley, *Inventions That Shaped the World: The Automobile* (New York: Scholastic, 2005), 5–41.

3. John Bankston, *Karl Benz and the Single Cylinder Engine* (Hockessin, Delaware: Mitchell Lane Publishers, 2005), 13–41.

4. Lindsay Brooke, *Ford Model T: The Car That Put the World on Wheels* (Minneapolis: MBI Publishing Company), 48–74.

5. World Business Council for Sustainable Development, "Facts and Trends: 2050," www.wbcsd.org/DocRoot/xxSdHDlXwf1J2J3ql0I6/Basic-Facts -Trends-2050.pdf.

6. "Methanol Fuels Indy Race Cars and Cleaner Air," Methanol Institute, May 21, 1997, www.methanol.org/altfuel/press/pr970521.html.

7. Andrew Leonard, "Why Is Diesel Even More Expensive Than Gas?" *Salon*, June 2008, www.salon.com/tech/htww/2008/06/02/why_is_diesel_so _expensive/index.html.

8. Karen Breslau, "A Natural Road Trip: Passing the 'Fossil Fools' in a CNG-powered car," *Newsweek*, August 22, 2008, www.newsweek.com/id/154709.

9. Maryland Energy Administration, "Straight Answers on Advanced Technologies: Propane," October 2006, www.energy.state.md.us/incentives/transportation/factsheets/Propane.pdf.

10. U.S. Energy Information Administration, "Propane Explained," http://tonto.eia.doe.gov/energyexplained/index.cfm?page=propane_home.

11. National Resource Council Committee, *Effectiveness and Impact of Corporate Average Fuel Economy (CAFE) Standards* (Washington, DC: National Academies Press, 2002), 13–30.

12. Bryan Walsh, "CAFE Standards: Fuzzy Math on Fuel Economy," *Time*, November 7, 2008, www.time.com/time/health/article/0,8599,1857620,00.html.

13. Stephen Power and Christopher Conkey, "U.S. Orders Stricter Fuel Goals for Autos," *Wall Street Journal*, May 19, 2009, http://online.wsj.com/article/SB124266939482331283.html.

14. U.S. Environmental Protection Agency, "2010 Fuel Economy Guide," www.fueleconomy.gov/feg/FEG2010.pdf.

Chapter 9: Biodiesel

1. National Biodiesel Board, "Frequently Asked Questions," www.biodiesel.org/resources/faqs/.

2. Wikipedia: The Free Encyclopedia, "Biodiesel," http://en.wikipedia.org/wiki/Biodiesel#Availability_and_prices.

3. Stefan Bringezu, Helmut Schutz, Meghan O'Brien, Lea Kauppi, Robert W. Howarth, and Jeff McNeely, "Assessing Biofuels," United Nations Environment Programme, www.unep.fr/scp/rpanel/pdf/Assessing_Biofuels_Full_Report.pdf.

4. John Sheehan, Terri Dunahay, John Benemann, and Paul Roessler, "A Look Back at the U.S. Department of Energy's Aquatic Species Program: Biodiesel from Algae," U.S. Department of Energy's Office of Fuels Development, July 1998, www.nrel.gov/docs/legosti/fy98/24190.pdf.

Chapter 10: Ethanol

1. Bill Moore, "Is Switch Grass Viable?" *EV World*, May 6, 2007, www.evworld.com/article.cfm?storyid=1246.

2. Howard Gruenspecht, "Testimony to U.S. House of Representatives," Subcommittee on General Farm Commodities and Risk Management, April 1, 2009, www.congressional.energy.gov/documents/4-1-09_Final_Testimony_(Gruenspecht).pdf.

3. Robert Rapier, "Gasoline Blending 101: The Ethanol Blending Requirement," The Oil Drum, July 28, 2008, www.theoildrum.com/node/4308.

4. "Renewable Fuel Standard Program," U.S. Environmental Protection Agency, November 19, 2009, www.epa.gov/OMS/renewablefuels/.

5. Rick Duke and Dan Lashof, "Putting America on the Path to Solving Global Warming," *NRDC Issue Paper*, June 2008, www.nrdc.org/globalWarming/energy/eeconomy.pdf.

6. Nathanael Greene, "Growing Energy: How Biofuels Can Help End America's Oil Dependence," National Resources Defense Council, December 2004, www.nrdc.org/air/energy/biofuels/biofuels.pdf.

7. Nathanael Greene and Yerina Mugica, "Bringing Biofuels to the Pump: An Aggressive Plan for Ending America's Oil Dependence," National Resource Defense Council, July 2005, http://bio.org/ind/advbio/NRDC.pdf.

8. John O. Christianson, "U.S. Ethanol Industry Efficiency Improvements: 2004 Through 2007," Christianson & Associates, PLLP, 2007.

9. Douglas Durante and Todd Sneller, "Net Energy Balance of Ethanol Production," Ethanol Across America, Spring 2009, www.cleanfuelsdc.org/pubs/documents/EnergyBalanceIssueBriefMarch09.pdf.

10. Keith Johnson, "More Bad News for Ethanol," *Wall Street Journal*, January 23, 2008, http://blogs.wsj.com/environmentalcapital/2008/01/23/more-bad-news-for-ethanol/.

11. Nathan Doyel, "The Costs and Benefits of Ethanol in the U.S. and Brazil," Georgetown University, 2007, www.outreachworld.org/Files/georgetown/EthanolLessonPlan.pdf.

Chapter 11: Hybrids, Plug-Ins, and Electric Vehicles

1. James Larminie and John Lowry, *Electric Vehicle Technology Explained* (United Kingdom: John Wiley & Sons, 2003), 1–7.

2. Air Resources Board, "Status Report on the California Air Resources Board's Zero Emission Vehicle Program," California Environmental Protection Agency, April 20, 2007, www.arb.ca.gov/msprog/zevprog/zevreview/zev_review_staffreport.pdf.

3. Mark Vaughn, "Electric Chic Gets Quicker," *Autoweek*, December 14, 2009, www.teslamotors.com/display_data/AutoWeek_Dec09.pdf.

4. Michael J. Scott, Michael Kintner-Meyer, Douglas B. Elliott, and William M. Warwick, "Impact Assessment of Plug-In Hybrid Vehicles on Electric Utilities and Regional U.S. Power Grids," November 2007, http://energytech.pnl.gov/publications/pdf/PHEV_Economic_Analysis_Part2_Final.pdf.

5. Electric Power Research Institute and National Resource Defense Council, "Environmental Assessment of Plug-in Hybrid Electric Vehicles," http://mydocs.epri.com/docs/public/PHEV-ExecSum-vol1.pdf.

Chapter 12: Fuel Cells and Hydrogen

1. Candace Lombardi, "Studying the Hydrogen Energy Chain," CNET News, April 4, 2007, http://news.cnet.com/Studying-the-hydrogen-energy-chain/2100-11392_3-6173003.html.

2. Robert Rose, "Questions and Answers about Hydrogen and Fuel Cells," Breakthrough Technologies Institute,   www.fuelcells.org/info/library/QuestionsandAnswers062404.pdf.

3. "Hydrogen Production," U.S. Energy Information Administration—Energy Efficiency and Renewable Energy, December 12, 2008, www1.eere.energy.gov/hydrogenandfuelcells/production/.

4. Otis Port, "Hydrogen Cars Are Almost Here, But . . ." *BusinessWeek*, January 24, 2005, www.businessweek.com/magazine/content/05_04/b3917097_mz018.htm.

Chapter 13: The Road Ahead

1. Energy Independence and Security Act of 2007, http://frwebgate.access.gpo.gov/cgi-bin/getdoc.cgi?dbname=110_cong_bills&docid=f:h6enr.txt.pdf.

2. Fred Sissine, "Energy Independence and Security Act of 2007: A Summary of Major Provisions," December 21, 2007, http://energy.senate.gov/public/_files/RL342941.pdf.

3. "Clean Pass–HOV," Department of Environmental Conservation, March 2006, www.dec.ny.gov/environmentdec/19011.html.

Chapter 14: Security, Reliability, Diversity, and Independence

1. International Energy Agency, "World Energy Outlook 2009," www.worldenergyoutlook.org.

2. World Economic Forum and Cambridge Energy Research Associates, "The New Energy Security Paradigm," Spring 2006, www.weforum.org/pdf/Energy.pdf.

3. U.S. Energy Information Administration, "World Oil Transit Chokepoints," January 2008, www.eia.doe.gov/cabs/World_Oil_Transit_Chokepoints/Background.html.

4. John Deutch, James R. Schlesinger, and David G. Victor, "National Security Consequences of U.S. Oil Dependency," Council on Foreign Relations, October 2006, http://www.cfr.org/content/publications/attachments/EnergyTFR.pdf.

5. Extractive Industries Transparency Initiative, "The EITI Principles and Criteria," www.eitransparency.org.

Chapter 15: Our Aging Power Grid

1. Earth Hour, www.earthhour.org.

2. U.S.–Canada Power System Outage Task Force, "Final Report on the August 14, 2003, Blackout in the United States and Canada: Causes and Recommendations," April 2004, https://reports.energy.gov/BlackoutFinal-Web.pdf.

3. Baltimore Gas & Electric, "What Causes Power Outages?" www.bge.com/portal/site/bge/menuitem.fe9c7e782b73e84606370f10d66166a0/.

4. U.S.–Canada Power System Outage Task Force, "Final Report on the August 14, 2003, Blackout in the United States and Canada: Causes and Recommendations," April 2004, https://reports.energy.gov/BlackoutFinal-Web.pdf.

5. M. Shahidehpour, "Our Aging Power Systems: Infrastructure and Life Extension Issues," *IEEE Power and Energy Magazine*, May–June 2006, http://ieeexplore.ieee.org/xpl/freeabs_all.jsp?tp=&arnumber=1632450&isnumber=34233.

6. Max Schulz, "Americans Consume Lots of Myths with Their Energy," Manhattan Institute for Policy Research, April 24, 2007, www.manhattan-institute.org/html/miarticle.htm?id=4063.

7. Electric Power Research Institute and the Electricity Innovation Institute, "The Cost of Power Disturbances to Industrial and Digital Economy Companies," June 2001, www.epri-intelligrid.com/intelligrid/docs/Cost_of_Power_Disturbances_to_Industrial_and_Digital_Technology_Companies.pdf.

8. Mark W. Chupka, Robert Earle, Peter Fox-Penner, and Ryan Hledik, "Transforming America's Power Industry: The Investment Challenge 2010–2030," The Edison Foundation, November 2008, www.eei.org/ourissues/finance/Documents/Transforming_Americas_Power_Industry.pdf.

9. Ben Block, "U.S. Government Seeks to Limit Federal Energy Use," August 21, 2009, www.worldchanging.com/archives/010376.html.

10. Kim Quillen, "Study Pinpoints the Cost of Upgrading the Electrical Transmission Grid," Nola.com, February 14, 2009, http://blog.nola.com/tpmoney/2009/02/study_pinpoints_the_cost_of_up.html.

Chapter 16: Nuclear Energy

1. Bill Willis, "E = MC² Explained," Worsley School Online, 2009, www.worsleyschool.net/science/files/emc2/emc2.html.

2. "Nuclear Weapon," *Encyclopedia Britannica*, 2009, www.britannica.com/EBchecked/topic/421827/nuclear-weapon/275625/Principles-of-atomic-fission-weapons.

3. James Mahaffey, *Atomic Awakening: A New Look at the History and Future of Nuclear Power* (New York: Pegasus, 2009), 47–63.

4. Helen Caldicott, *Nuclear Power Is Not the Answer* (New York: The New Press, 2006).

5. Office of Civilian Radioactive Waste Management, "The National Repository at Yucca Mountain," U.S. Department of Energy, July 2008, www.ocrwm

.doe.gov/uploads/1/06237PD_The_National_Repository_at_Yucca_Mountain .pdf.

6. David Bodansky, "Comments on Yucca Mountain and Nuclear Energy," American Physical Society, April 2009, www.aps.org/units/fps/newsletters/ 200904/bodansky.cfm.

7. Dan Shapley, "Rock Stars vs. Nuclear Power," *The Daily Green*, October 23, 2007, www.thedailygreen.com/environmental-news/latest/Rock-Stars-Nuclear -Power-47102307.

8. Lisa Goff, "Quick Study: The Facts on Nuclear Energy," *Reader's Digest*, August 2008, www.rd.com/your-america-inspiring-people-and-stories/nuclear -energy-facts/article81880.html.

9. Kenneth Stier, "Nuclear: What Is It and How Does It Work?" CNBC, June 20, 2008, www.cnbc.com/id/24875745.

10. "Nuclear Europe: Country Guide," BBC News, April 15, 2009, http:// news.bbc.co.uk/2/hi/europe/4713398.stm.

11. Stephen J. Dubner and Steven D. Levitt, "The Jane Fonda Effect," *New York Times*, September 16, 2007, www.nytimes.com/2007/09/16/magazine/16wwln -freakonomics-t.html.

Chapter 17: The Geopolitics of Oil

1. Daniel Yergin, *The Prize: The Epic Quest for Oil, Money and Power* (New York: Free Press, 2008).

2. Sonia Shah, *Crude: The Story of Oil* (New York: Seven Stories Press, 2004), 17–49.

3. Mark Trumbull, "Risk of Rising Oil Nationalism," *Christian Science Monitor*, April 3, 2007, www.csmonitor.com/2007/0403/p04s01-usec.html.

Chapter 18: Reducing Energy Wastage

1. United Nations Department of Economic and Social Affairs, United Nations Development Programme, and World Energy Council, *World Energy Assessment: Energy and the Challenge of Sustainability* (United Nations Development Programme, World Energy Council Press, 2000), chapter 6, 174–216.

2. Donald R. Wulfinghoff, *The Modern History of Energy Conservation: An Overview for Information Professionals* (Wheaton, Maryland: Energy Institute Press, 2003), 2–8.

3. Jay E. Hakes, "Administrator's Message: 25th Anniversary of the 1973 Oil Embargo," U.S. Energy Information Administration, September 3, 1998, www .eia.doe.gov/emeu/25opec/anniversary.html.

Chapter 19: Energy Efficiency Begins at Home

1. Penni McLean-Conner, *Energy Efficiency: Principles and Practices* (Tulsa, Oklahoma: PennWell Books, 2009), 71–113.

2. Donald R. Wulfinghoff, *Energy Efficiency Manual* (Wheaton, Maryland: Energy Institute Press, 1999), 1199–1239.

3. John Krigger and Chris Dorsi, *The Homeowner's Handbook to Energy Efficiency* (Helena, Montana: Saturn Resource Management, 2008), 3–27.

4. Robert Lamb, "How Vampire Power Works," HowStuffWorks, http://electronics.howstuffworks.com/gadgets/other-gadgets/vampire-power3.htm.

5. "Kill A Watt™ PS," P3 International, 2008, www.p3international.com/products/consumer/p4320.html.

6. Community Environmental Council, "Resources to Green Your Home," www.cecsb.org/index.php?option=com_content&task=view&id=64&Itemid=103.

7. U.S. Environmental Protection Agency, "About ENERGY STAR," U.S. Department of Energy, www.energystar.gov/index.cfm?c=about.ab_index.

8. U.S. Department of Energy, "How to Read the EnergyGuide Label," June 2, 2006, www1.eere.energy.gov/consumer/tips/energyguide.html.

9. *Water Conservation* (Irvine, California: Saddleback Educational Publishing, 2009).

10. U.S. Department of Energy, "Energy Savers: Tips on Saving Energy and Money at Home," Energy Efficiency and Renewable Energy, April 9, 2009, www1.eere.energy.gov/consumer/tips/lighting.html.

11. GE Lighting, "Compact Fluorescent Light Bulb (CFL) FAQs," General Electric Company, 2009, www.gelighting.com/na/home_lighting/ask_us/faq_compact.htm.

12. U.S. Department of Energy, "Information on Compact Fluorescent Light Bulbs (CFLs) and Mercury," Energy Efficiency and Renewable Energy, July 2008, www.energystar.gov/ia/partners/promotions/change_light/downloads/Fact_Sheet_Mercury.pdf.

Chapter 20: Green Living

1. Peter Jones, Jan Bebbington, Geoffrey Boulton, Martyn Evans, Campbell Gemmell, Nick Hanley, Patrick Harvie, George Hazel, Iain McMillan, Ian Marchant, Michael Northcott, Simon Pepper, Susan Roaf, Scottish Youth Parliament, Jim Skea, Richard Wakeford, and David C. Watt, "Reducing Carbon Emissions—The View from 2050," October 14, 2008, www.jmt.org/assets/john%20muir%20award/downloads/hop%2079%20reducing%20carbon%20emissions%20-%20the%20view%20from%202050.pdf.

2. Yvonne Jeffery, Liz Barclay, and Michael Grosvenor, *Green Living for Dummies* (Hoboken, New Jersey: Wiley Publishing, 2008), 1–6.

3. U.S. Green Building Council, "An Introduction to LEED," www.usgbc.org/DisplayPage.aspx?CategoryID=19.

4. Jon Creyts, Anton Derkach, Scott Nyquist, Ken Ostrowski, and Jack Stephenson, "Reducing U.S. Greenhouse Gas Emissions: How Much at What Cost?" McKinsey & Company, December 2007, www.mckinsey.com/clientservice/ccsi/pdf/US_ghg_final_report.pdf.

5. Paul McArdle, "Emissions of Greenhouse Gases in the United States 2008," U.S. Department of Energy, December 2009, ftp://ftp.eia.doe.gov/pub/oiaf/1605/cdrom/pdf/ggrpt/057308.pdf.

Chapter 21: Energy Efficiency on a Broader Scale

1. Bob Aldrich, "Saving Time, Saving Energy: Daylight Saving Time—It's History and Why We Use It," April 21, 2009, www.energy.ca.gov/daylightsaving.html.

2. Olivia Zaleski, "Environmental News: Big Media Catches On—Major TV Networks Catch Green Fever," *The Daily Green*, October 23, 2007, www.thedailygreen.com/environmental-news/latest/TV-Goes-Green-47102308.

3. Allison Linn, "Corporations Find Business Case for Going Green," MSNBC, April 18, 2007, www.msnbc.msn.com/id/17969124/.

Chapter 22: Cleantech Investments

1. California Natural Resources Agency, "The California Gold Rush," State of California, http://ceres.ca.gov/ceres/calweb/geology/goldrush.html.

2. "The California Gold Rush, 1849," EyeWitness to History, 2003, www.eyewitnesstohistory.com/californiagoldrush.htm.

3. Steven Milunovich, "The Sixth Revolution: The Coming of Cleantech," *Merrill Lynch—Industry Overview,* November 17, 2008, www.responsible-investor.com/images/uploads/resources/research/21228316156Merril_Lynch-_the_coming_of_clean_tech.pdf.

4. Charles Morand, "Cleantech, Optimism Squared and the Family Reunion Investment Test," Alt Energy Stocks, March 31, 2009, www.altenergystocks.com/archives/2009/03/cleantech_optimism_squared_and_the_family_reunion_investment_test.html.

5. Neal Dikeman, "What Is Clean Tech?" CNET News, August 10, 2008, http://news.cnet.com/8301-11128_3-10012950-54.html.

6. Tracy T. Lefteroff, Fred Sroka, Greg Vlahos, and Belanne Ungarelli, "The Exit Slowdown and the New Venture Capital Landscape," *The MoneyTree Report*, Sep-

tember 2008, PriceWaterhouseCoopers—www.pwcmoneytree.com/MTPublic/ns/moneytree/filesource/exhibits/The%20exit%20slowdown%20and%20the%20new%20venture%20capital%20landscape.pdf.

7. United Nations Environment Programme, "Global Trends in Sustainable Energy Investment 2009," http://sefi.unep.org/fileadmin/media/sefi/docs/publications/Executive_Summary_2009_EN.pdf.

8. "Wall Street's New Love Affair—Special Report," *Business Week*, August 14, 2006, www.businessweek.com/magazine/content/06_33/b3997073.htm?chan=top+news_top+news.

9. World Energy Council, "The Energy Industry Unveils Its Blueprint for Tackling Climate Change," March 2007, www.worldenergy.org/documents/stat2007.pdf.

10. Gerald Doucet, "The Energy Industry Unveils Its Blueprint for Tackling Climate Change," World Energy Council, March 22, 2007, www.worldenergy.org/news__events/media_relations/press_releases/184.asp.

11. Ron Pernick, Clint Wilder, Dexter Gauntlett, and Trevor Winnie, "Clean Tech Job Trends 2009," Clean Edge, October 2009, www.cleanedge.com/reports/pdf/JobTrends2009.pdf.

12. Lori Valigra, "Cleantech Eclipses IT and Biotech as the Investors' Favorites," Science-Business Network, October 15, 2009, http://bulletin.sciencebusiness.net/ebulletins/showissue.php3?page=/548/art/15496.

13. Michael Liebreich, Chris Greenwood, Max von Bismarck, and Anuradha Gurung, "Green Investing—Towards a Clean Energy Infrastructure," World Economic Forum, January 2009, www.newenergymatters.com/UserFiles/File/WEF_NEF_2009_01_30_Davos_Green_Investing_Report.pdf.

14. Brent D. Yacobucci and Randy Schnepf, "Ethanol and Biofuels: Agriculture, Infrastructure, and Market Constraints Related to Expanded Production," *CRS Report for Congress*, March 16, 2007, http://collinpeterson.house.gov/PDF/ethanol.pdf.

15. "The Global Dynamics of Biofuels: Potential Supply and Demand for Ethanol and Biodiesel in the Coming Decade," Brazil Institute Special Report, Woodrow Wilson International Center for Scholars, April 2007, www.wilsoncenter.org/topics/pubs/Brazil_SR_e3.pdf.

16. Robert Kunzig, "The Big Idea," *National Geographic*, December 2009, http://ngm.nationalgeographic.com/big-idea/05/carbon-bath.

17. www.recovery.gov, U.S. government's official website providing data related to Recovery Act spending.

18. Sue Kirchhoff, "How Will the $787 Billion Stimulus Package Affect You?" *USA Today*, February 17, 2009, www.usatoday.com/money/economy/2009-02-12-stimulus-package-effects_N.htm.

19. John Downey, "Duke Energy Teams with China Huaneng on Clean Energy," *Charlotte Business Journal*, August 10, 2009, www2.nccommerce.com/eclipsfiles/20900.pdf.

Chapter 23: Energy, Economy, Jobs, and Education

1. Bureau of Labor Statistics, "Employment Situation Summary," U.S. Department of Labor, December 4, 2009, www.bls.gov/news.release/empsit.nr0 .htm.

2. Testimony of David Foster, Executive Director of the Blue Green Alliance, to the Committee on Energy and Commerce and the Subcommittee on Energy and Environment Hearing on the American Clean Energy and Security Act of 2009, April 22, 2009, www.bluegreenalliance.org/admin/publications/ files/0008.4.pdf.

3. Robert Polli, Heidi Garrett-Peltier, James Heintz, and Helen Scharber, "Green Recovery: A Program to Create Good Jobs and Start Building a Low-Carbon Economy," Center for American Progress, September 2008, www.americanprogress .org/issues/2008/09/pdf/green_recovery.pdf.

Chapter 24: The Coming Energy Revolution

1. "An Introduction to LEED," U.S. Green Building Council, 2009, www .usgbc.org/DisplayPage.aspx?CategoryID=19.

2. Anya Kamenetz, "The Green Standard?" *Fast Company*, December 19, 2007, www.fastcompany.com/magazine/119/the-green-standard.html.

3. Andrew C. Burr, "CoStar Study Finds Energy Star, LEED Buildings, Outperform Peers," CoStar Group, March 26, 2008, www.costar.com/News/Article .aspx?id=D968F1E0DCF73712B03A099E0E99C679.

4. United States Climate Action Partnership, www.us-cap.org.

5. Climate Savers Computing, www.climatesaverscomputing.org.

Selected Bibliography
Volume 2: Alternative Energy

Bankston, John. *Karl Benz and the Single Cylinder Engine. Uncharted, Unexplored, and Unexplained* (Hockessin, Delaware: Mitchell Lane Publishers, 2004).

Barclay, Liz, Michael Grosvenor, and Yvonne Jeffery. *Green Living for Dummies* (Hoboken, New Jersey: Wiley Publishing, 2008).

Bates, Diana, Karl Gawell, and Alyssa Kagel. *A Guide to Geothermal Energy and the Environment*. Geothermal Energy Association, 2007. Online at http://www. geo-energy.org/publications/reports/Environmental%20Guide.pdf.

Baxter, Richard. *Energy Storage: A Nontechnical Guide* (Tulsa, Oklahoma: PennWell Books, 2005).

Brockey, Liam Matthew. *Journey to the East: The Jesuit Mission to China 1579–1724* (Cambridge, Massachusetts: Harvard University Press, 2007).

Brooke, Lindsay, Bill Ford, and Patricia Mooradian. *Ford Model T: The Car That Put the World on Wheels* (Minneapolis: MBI Publishing Company, 2008).

Caldicott, Helen. *Nuclear Power Is Not the Answer* (New York: The New Press, 2006).

Committee on Materials Research for Defense After Next, and National Research Council. *Materials Research to Meet 21st-Century Defense Needs* (Washington, DC: National Academies Press, 2003).

Conley, Robyn. *The Automobile. Inventions That Shaped the World* (New York: Scholastic, 2005).

Dorsi, Chris, and John Krigger. *The Homeowner's Handbook to Energy Efficiency* (Helena, Montana: Saturn Resource Management, 2008).

Larminie, James, and John Lowry. *Electric Vehicle Technology Explained* (United Kingdom: John Wiley & Sons, 2003).

Mahaffey, James. *Atomic Awakening: A New Look at the History and Future of Nuclear Power* (New York: Pegasus Books, 2009).

McLean-Conner, Penni. *Energy Efficiency: Principles and Practices* (Tulsa, Oklahoma: PennWell Books, 2009).

National Research Council. *Effectiveness and Impact of Corporate Average Fuel Economy (CAFE) Standards* (Washington, DC: National Academies Press, 2003).

Shah, Sonia. *Crude: The Story of Oil* (New York: Seven Stories Press, 2004).

United Nations Department of Economic and Social Affairs, United Nations Development Programme, and World Energy Council. *World Energy Assessment: Energy and the Challenge of Sustainability* (New York: United Nations Development Programme, 2000). Online at http://www.undp.org/energy/activities/wea/drafts-frame.html.

Water Conservation (Irvine, California: Saddleback Educational Publishing, 2009).

Wulfinghoff, Donald R. *Energy Efficiency Manual* (Wheaton, Maryland: Energy Institute Press, 1999).

————. *The Modern History of Energy Conservation: An Overview for Information Professionals* (Wheaton, Maryland: Energy Institute Press, 2003). Online at http://www.energybooks.com/resources/modern_history_of_energy.pdf

Yergin, Daniel. *The Prize: The Epic Quest for Oil, Money and Power* (New York: Free Press, 2008).

Index
Volume 2: Alternative Energy

About the Authors

Vikram Janardhan is chief executive officer of Insera Energy LLC, a consulting firm providing services to electric utilities and power generation companies. Until 2007, he was the president of Global Energy Software, an industry-leading provider of strategic consulting and software to energy organizations worldwide. In his spare time he oil paints, practices yoga, and takes far too many photos of his nineteen-month-old daughter. He lives with his wife and daughter in Sacramento, California.

Bob Fesmire is a writer and communications manager with over twelve years of experience in the energy industry. Presently he is responsible for media relations for ABB in North America, a multi-billion-dollar business spanning automation and power technologies. Bob is also an avid gardener, lifelong ice hockey player, and interested follower of events in Washington, DC. He and his wife Gina relocated from Sunnyvale, California to Raleigh, North Carolina, in the summer of 2009.